Site Investigation, Remediation, and Closure

A Simplified Guide for Environmental and Real Estate Professionals

Arshud Mahmood, PhD, PE
Principal Engineer/President
MCE Group

Part II: Remediation
prepared in collaboration with
Mohammad Estiri, PhD
Consulting Environmental Engineer

Government Institutes
Rockville, MD

Government Institutes, Inc., 4 Research Place,
Rockville, Maryland 20850, USA.
Phone: (301) 921-2300
Fax: (301) 921-0373
Email: giinfo@govinst.com
Internet address: http://www.govinst.com

02 01 00 99 5 4 3 2

The figures on pages 21, 100, and 101 appeared in "Moisture and Vapor Extraction Don't Mix," *Soil & Groundwater Cleanup*, March 1996, and are reprinted with permission.

The case history on page 138 was provided by Treadwell & Rollo, Inc., San Francisco, CA.

Library of Congress Cataloging-in-Publication Data

Mahmood, Arshud.
Site investigation, remediation, and closure : a simplified guide for environmental and real estate professionals / by Arshud Mahmood.
p. cm.
Includes index.
ISBN: 0-86587-623-1
1. Hazardous waste sites--Management. I. Title.
TD1052.M35 1998
628.5--dc21
98-16429
CIP

Printed in the United States of America

for my parents

Summary Table of Contents

CONTENTS

PART II –SITE REMEDIATION

PART III – SITE CLOSURE

List of Figures and Tables

Figures

List of Tables

Author Biography

Arshud Mahmood obtained his secondary education in Quetta, Baluchistan. He then completed a bachelor's degree in physics at Government College, Lahore, and a bachelor's degree in civil engineering at University of Karachi, Pakistan. His graduate work began at Asian Institute of Technology, Bangkok, Thailand, with a master's degree in soil engineering. The graduate studies continued at University of California, Berkeley, where he completed master's and doctorate degrees in geotechnical engineering. He has taught civil and environmental engineering courses at California State University, Los Angeles, and at University of California, Los Angeles.

Dr. Mahmood has spent the last twenty-five years in consulting assignments with leading engineering and environmental firms including Woodward-Clyde Consultants and Dames & Moore. During the last fifteen years he has been working on the problems of construction at difficult sites, including contaminated sites. On a number of occasions he has served as a designated expert on projects with construction defects or inadequate remediation. He was awarded the Engineering Excellence Award from the Consulting Engineers Association of California for innovative use of liner protection for subterranean construction at a contaminated site. Projects on which he served as a project manager, task leader or senior consultant have included design and construction of commercial and office buildings, shopping malls, industrial facilities, water and wastewater treatment plants, transportation facilities, and landfill closures.

He holds registrations as a civil engineer in Alaska, California, Nevada, and Texas. He is also a registered geotechnical engineer and a registered environmental assessor in California. He is a current member of American Society of Testing and Materials, American Society of Civil Engineers and American Concrete Institute. He currently heads MCE Group, a geotechnical and environmental engineering consulting firm located in California.

ACKNOWLEDGMENTS

During my education and professional career, I have received guidance, encouragement and advice from teachers, colleagues and clients. It is impossible to now separate the contributions of others from my own ideas. I want to acknowledge those who provided me opportunities for creative development: Professor N. Ahmad, Professor R.E. Mirza, Professor Za-Chieh Moh, Professor James K. Mitchell, Dr. Robert L. McNeill, Mr. Herman Seick, Mr. Bram McClelland, Mr. Art Darrow, Joyia Greenfield, Esq., and Bruce Lorber, Esq.

Over the years a number of colleagues and clients provided friendly guidance, including: M. Ismail, M.H. Dawood, Dai Watkins, Don Treadwell, Hal Meadow, Larry Voors, and Linda Harper Wisener, Esq.

Several people have made direct contributions to the preparation of this text, including students who sat through my lectures at University of California, Los Angeles. Dr. Mohammad Estiri, Consulting Engineer, was my collaborator in the preparation of Part II on Remediation. He provided many original ideas and technical details that have become incorporated in the test. Tanweer Shah and Brian Wynne assisted me at Dames & Moore, as we jointly developed solutions for design and construction on contaminated sites. Roy Patterson reviewed Chapters 1 and 2. Lita Navarro typed, formatted and arranged the text, and made many helpful suggestions during my endless revisions. Ed King did the artwork for illustrations, and Jim Beardsley assisted in bibliography compilation

MCE Group, my present firm, allowed me sufficient time and resources to transform lecture material into the text you are about to read. The firm is a nascent enterprise, offering services in geotechnical engineering and environmental consulting. I wish to acknowledge the support of Houman Makarechi, whose brainchild MCE Group was. Support from Jamshid Amanat and Farid Motamed was also valuable.

My wife, Maliha, believed all along that I could complete this book, and our son Omar and our daughter Alia reinforced this affirmation. My brother Parvez provided encouragement and support.

Publication Credits: Soil and Groundwater Cleanup granted permission to reproduce five illustrations that first appeared in the magazine. Treadwell & Rollo provided the case history on brownfields.

INTRODUCTION

It has happened many times. A vacant lot is scheduled to become the site of an office complex. Designs are completed, a leasing office opens, the developer ceremoniously shovels the first spade of dirt as bulldozers and backhoes start their engines. The project begins with a roar.

Everything looks great at the start of excavation, then oily soils are discovered a week later and the entire project grinds to a halt. A belated site assessment finds that an old gas station, a factory, or a manufactured gas plant once occupied the area, and the headaches begin. In many industrialized, urban centers of the world, as many as one-third of the construction sites may be afflicted with soil contamination. Fuel hydrocarbons, metals, and solvents are the most common contaminants. In agricultural areas, suburbs carved out of former farmlands often encounter pesticides and herbicides. Urban expansion into what used to be farmland, including land that was used to warehouse and handle agricultural chemicals, has uncovered contaminated land.

How to best take care of contaminated land and associated water is the subject of this text. In the past, expensive cleanups have been driven by one overarching force—the hazardous waste regulations. Alternately, adventurous souls have dug up the wastes thoughtlessly, only to widen the spread of contamination. Neither approach is acceptable any more, as society demands responsible yet inexpensive solutions from the environmental professionals. A lot of the resources have gone into investigations and litigation, and not enough into cleanup of contaminated sites. The earlier question of "How clean is clean" has been replaced by "How risky is risky", leading to the use of health-risk based clean-up criteria. This approach has dramatically increased the number of sites eligible for closure, without having to remove every last particle of chemicals from the site. The grouping of sites into Contaminated and Clean is no longer valid; we must think of a few properties as Impacted by contamination, and the rest that are not impacted. If the health risk is acceptable, then some amount of chemicals can be left at the site safely and without violating any regulations.

Brownfields provide us with ample opportunities for application of a "start-to-finish" approach to management of contaminated properties. Land developers have avoided distressed industrial properties in the inner cities and other urban settings because of a fear of lingering environmental liabilities. Their pessimistic evaluations are based on the premise that a contaminated industrial property that is surrounded by ongoing industrial operations has to be cleaned up to the same standards as a pristine meadow in the wilderness, or as completely as a kindergarten school in a residential neighborhood. Contrarily, brownfields require cleanup only to the relatively high background concentrations representative of their setting. Factors such as proximity to other industrial operations and the lack of sensitive receptors are considered, making brownfields viable for redevelopment.

Remediation of wastes at brownfields construction sites require a mindset different from the cleanup-at-any-cost mentality. It is only necessary to disturb wastes in the zone that will be used by the proposed construction. It will, however, be necessary to protect the constructed facilities from the effects of whatever waste is left in the ground. This way, the future occupants of the buildings are protected and the building owners do not have to incur any unnecessary expense.

If contamination is discovered at a construction site, it should be removed without endangering the health and safety of the site workers or the surrounding community. However, the site does not have to turn into a hazardous waste zone. With careful monitoring, flexible boundaries can be set up for the hot zone, entry into which may require adherence to a site-specific health and safety plan. The remainder of the site can then be operated as an ordinary construction site.

The following text is divided into three parts, each covering a major subject area needed for managing contaminated properties. *Part I, Site Investigation*, imparts to the reader the theoretical and practical knowledge regarding the methodology of site investigation in three chapters. Chapter 1 discusses how to plan a site investigation while balancing the financial, regulatory, and scientific requirements. Chapter 2 on Environmental Setting deals with the subsurface character of the soil, rock, and groundwater at a site. Chapter 3 describes investigative techniques including Phase I and Phase II Assessments.

Part II, Remediation, covers the methods available for treating contaminated sites. This part of the book is divided into four chapters: Chapter 4 describes remedial alternative selection for a site, followed by Chapters 5, 6 and 7 that provide details of the remediation methods. Descriptions of the In-Situ Remediation Methods are presented in Chapter 5, Ex-Situ Methods in Chapter 6 and the Innovative Techniques in Chapter 7.

Part III, Closure, goes beyond site investigation methodology and remedial techniques, and covers four topics that are important for successful completion of a site closure. These are: Management of Environmental Data, including GIS, modeling, visualization, and other applications (Chapter 8); Health-Risk Based Assessments (Chapter 9), Regulatory Framework (Chapter 10), and, Confirmation Sampling and Site Closure (Chapter 11), including the use of statistical decision-making tools. Chapters 9 and 11 can be used to select acceptable remediation alternatives for brownfields, while keeping the costs reasonable.

This book presents information that will allow the reader to make decisions for contaminated sites. The intent here is to present information that allows the reader to handle waste within acceptable health risk and proceed with construction or project development at a reasonable cost.

PART I

SITE INVESTIGATION

CHAPTER 1

PLANNING A SITE INVESTIGATION

1.1 SETTING THE STAGE

Objectives

There are two main objectives of an investigation: (1) assess site hazard or risk, and (2) evaluate the feasibility of remediation. Each investigation program must fulfill these two objectives in accordance with the criteria described below.

Criteria

A site investigation must simultaneously fulfill three types of criteria: engineering, regulatory and financial.

The engineering criteria require that an appropriate confidence level be achieved in the amount and accuracy of the data obtained. Geostatistics are used to calculate whether the distribution of samples and the range of concentrations provide us with a high enough confidence level for design. The analytical test results of contaminants and measurements of the physical properties of site media are then used to design the remedial solution.

The regulatory criteria for defining a plume usually require that site investigations continue until nondetectable concentrations are found at the edges of the contaminated areas. This is not always achievable due to the presence of buildings and other facilities, and contaminant migration into adjacent properties.

The financial criteria, including project schedule, also have to be satisfied, and the work completed within the budget available and the time allotted. Every project manager has to fulfill this triad of engineering, regulatory, and financial requirements, and the challenge is to stay within the triangle's boundary. The exact scope of the investigation may vary over a broad range in accordance with the factors described below.

Scope

The scope of a site investigation is governed by the following five items:

- Geologic Complexity
- Contaminant Diversity and Distribution
- Regulatory Requirements
- Remedial Implications
- Budget, Schedule and Communications Strategy

Each of the above items is discussed separately in the following sections.

1.2 GEOLOGIC COMPLEXITY

Greater geologic complexity at a site increases the paths available for contaminant transport and dispersion, thus requiring greater effort in characterization and remediation. Complex stratigraphy in the subsurface can result from a number of geologic mechanisms, some of which are:

- Multiple formational origins for rock and soil
- Sedimentary depositional mechanisms and resulting lateral and vertical variations in material properties
- Secondary, post-depositional processes

An example of the first mechanism is when distinct formational processes—alluvial, glacial or aeolian—occur one after the other at a site. Soil and rock materials that originated by the action of glaciers may then be eroded—and redeposited—by a stream or river. This may lead to grain shapes and materials typical of glacial origin, and bedding and layering characteristics of water-borne sediments.

The second condition can result when river-deposited sediments are scoured away and then redeposited by stream action, leading to very finely bedded and criss-crossed alluvial layering. The influence of post-depositional folding, faulting, metamorphism or geochemical processes such as cementation can lead to the third condition.

Geologic complexity caused by any of the processes will require a greater investigative effort for adequate characterization. The number of borings required to define the stratigraphy at a complex site will be larger. Many of the strata will not be continuous, and the number of layers encountered in each boring may also be larger. The discontinuous layers will require intermediate borings.

Even if the layering is known in detail, the delineation of a contaminated zone is difficult at geologically complex sites. Highly mobile volatile liquids with a low viscosity can flow preferentially along bedding planes and spread in unanticipated directions. Large amounts of these contaminants can accumulate in finer-grained materials like silts, and are not easy to recover. Porous but discontinuous zones can act as reservoirs by accumulating viscous, low mobility contaminants that are very difficult to detect and remediate.

At some sites the soil materials have been deposited in one massive, homogeneous unit, with little or no layering or change in properties. At other sites the layering is horizontal, so that similar stratigraphy is encountered in every boring. Both types of sites would be considered to represent simple stratigraphy.

An example of complex stratigraphy is a site in the vicinity of downtown Los Angeles where excavation to a depth of 75 feet is planned. The site has nearly vertical bedding layers of alternating sandstone and siltstone as illustrated on Figure 1-1. Slightly contaminated water is expected in the excavation, but dewatering by predrainage is not feasible due to the obvious hydraulic discontinuities. Water level in adjacent borings has been measured at elevations that differ by as much as 40 feet, making an accurate investigation or remedial dewatering very difficult. It was finally decided to handle the site water within the excavation by using collection ditches across sandstone layers.

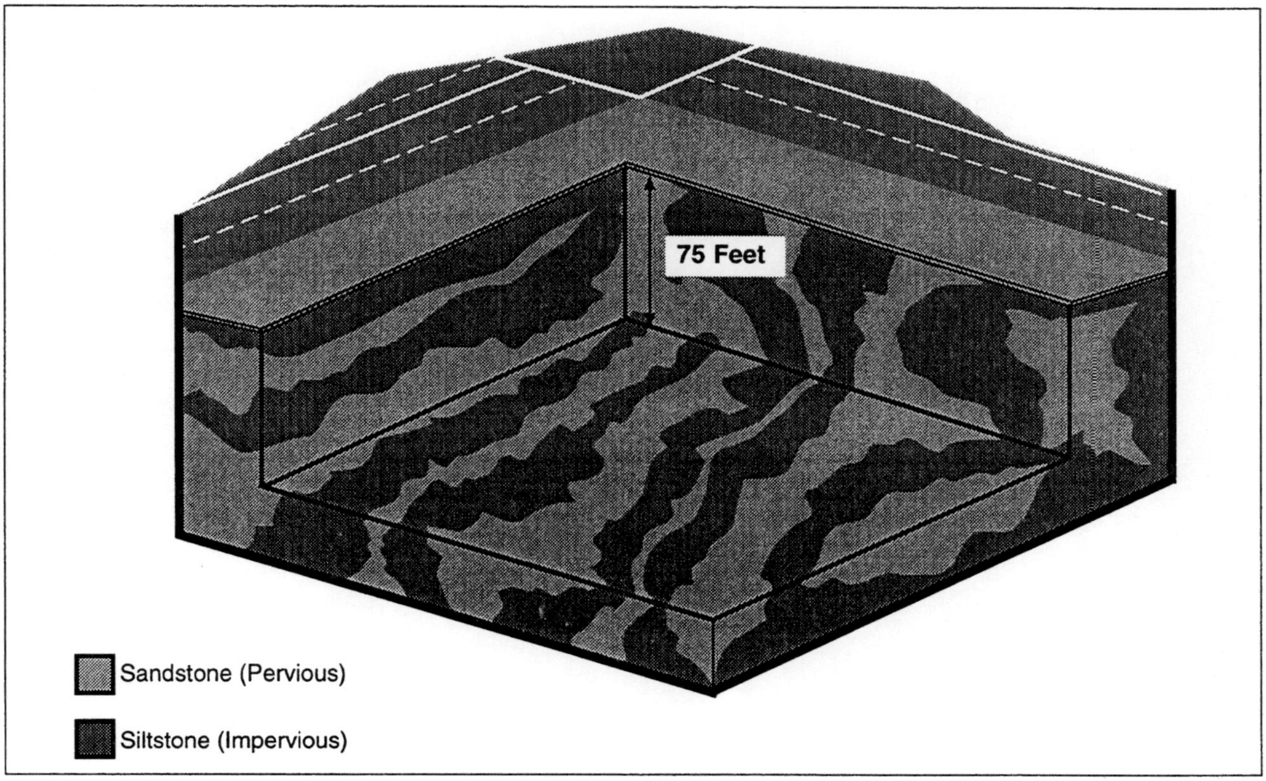

Figure 1-1. Complex Stratigraphy; Excavation Near Downtown Los Angeles

1.3 CONTAMINANT DIVERSITY AND DISTRIBUTION

A site with one or two similar contaminants (for instance, leaded gasoline and diesel fuel) and with a simple leakage scenario (point-source leak from an underground fuel tank) has a predictable plume that can be investigated in sufficient detail by using a small number of soil borings to a moderate depth. In contrast, sites with several different point sources, such as an industrial or processing facility with multiple product lines, clarifiers, and degreasers, may require a large number of borings, both shallow and deep.

A few general guidelines are stated below for site investigations for the four major contaminant groups.

Volatile Organic Compounds and Fuel Hydrocarbons

This group includes solvents (TCE, PCE), gasoline, diesel, and jet fuel. The presence of contaminants from this group is easy to identify because of their volatility—a cursory soil gas survey in any industrial neighborhood will turn up evidence of the presence of some solvent and fuel vapors in the shallow subsurface. The difficulty is in determining the significance of any vapor hits and in locating the source area, as measurements may represent stray vapors from adjacent properties, or may be remnants of past site uses, but no longer representative of contaminants in the soil.

Semivolatiles and Heavy Hydrocarbons

One of the key difficulties in characterizing semivolatile organic compounds is their high detection limit. In cases of co-contamination between PAHs and other hydrocarbons, detection limits may be so high as to prevent accurate risk assessment.

Typical PAHs include Benzo (a) Pyrene, and Anthracene, both of which have high viscosity and low mobility. No. 2 heating oil and furnace oil have similar mobility. In the vicinity of present or former oil fields, remnant crude oil may also be encountered. During investigation, the main pitfall to watch out for, is missing a subsurface accumulation or pool in a locally porous layer or a fracture zone. Otherwise, delineation of the contaminated area is generally not difficult. The other challenge is to identify and measure properties that will be required for remediation design.

Example

An urban brownfields property was the former site of a manufactured gas plant that contained waste oils and lamp black. These low mobility wastes were found to be limited to the upper 4 to 5 feet (Figure 1-2). The cleanup consisted of a shallow, site-wide excavation.

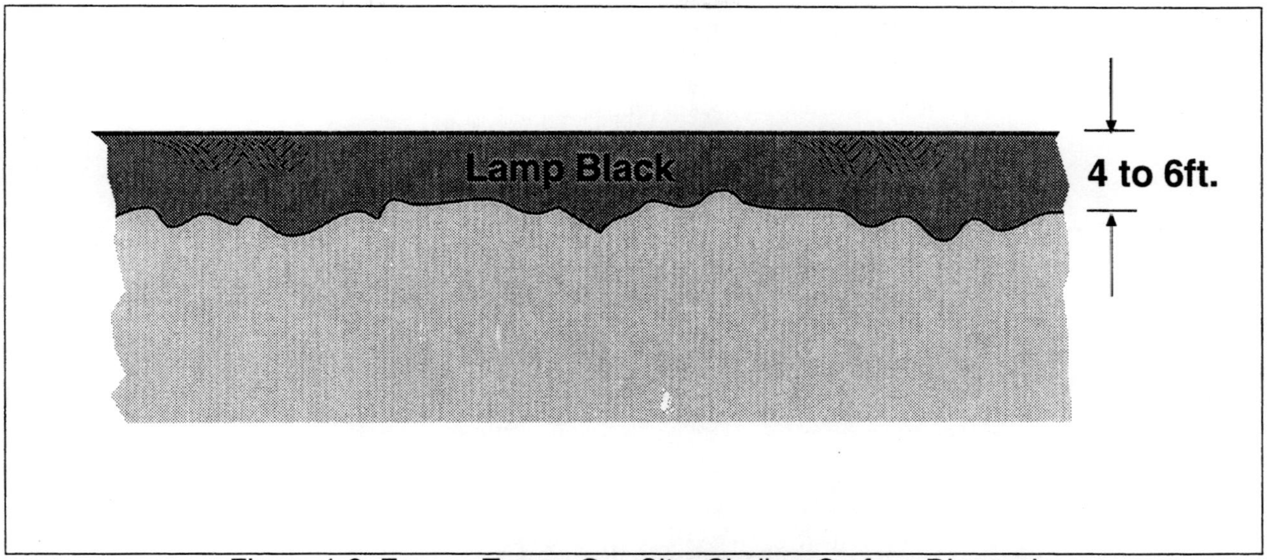

Figure 1-2. Former Towne Gas Site: Shallow Surface Disposal

Pesticides and PCBs

High concentrations of either pesticides or PCBs are generally localized and surficial, resulting from storage for farm use, and transformer leakage, respectively. Former farmland or orchards may contain widespread detectable concentrations of pesticides in the surficial soils. Sampling strategies for these surficial, often localized, and generally not very mobile, contaminants are limited to surface sampling or hand augering, and rarely include deep soil borings. Remediation involves shallow "scraping" excavations.

Example

An agricultural site that was being purchased for conversion to commercial use was known to contain pesticides and herbicides in the upper few inches. Drilling and sampling using a drill rig was considered inapplicable. Sampling strategy for the 100-acre site employed geostatistics, and the site was divided into thirty gridblocks of about 3 acres each. Based on random sampling procedures, six of the gridblocks were selected for detailed sampling. Surface-scraped samples were gathered in the six randomly selected grid blocks and the results provided a sufficiently high confidence level (Figure 1-3).

Figure 1-3. Random Sampling Strategy for Surficial Pesticide Contamination at a 100 Acre Site

Metals, Asbestos and Other Inorganic Wastes

The incidence of these materials in the subsurface soils is often related to past handling and disposal practices. These can be encountered in the subsurface either due to natural occurrence—generally in low concentrations on a regional basis, or due to past handling—in high but localized concentrations. For metals, the soluble compounds are much more toxic and thus hazardous, and typical background concentrations in soils found in industrial areas can produce cancer risks that exceed deminimis levels (e.g., arsenic).

1.4 REGULATORY REQUIREMENTS

Regulatory requirements influence scope of a site investigation. Underground Storage Tank (UST) investigations performed under fire department oversight are simple while a Superfund investigation for a large facility is performed in a much more formal manner. If, as part of a Superfund investigation, a UST is also discovered, often it will be handled in a much more detailed fashion, than if it was at an isolated service station.

A hypothetical comparison of a leaking UST discovered at four different locations that are being investigated in accordance with different regulations is as follows:

Leaking Underground Storage Tank Investigation				
Agency	*City Fire Department*	*County Health Agency*	*State Water Board*	*Federal EPA*
Facility	Steam Plant	Bank Site (Former Service Station)	Refinery (Brownfield)	Chemical Plant
Scope	2 borings to 40 ft	6 borings -10 to 60 ft	4 borings to 50 ft	6 borings to 50 ft

1.5 REMEDIAL IMPLICATIONS

At some sites the remedial options may become obvious very early in an investigation. Items learned during an investigation—site stratigraphy or contaminant distribution—may govern the remediation method. Other physical features—presence of buildings or the character of the surrounding area—may also limit the remedial choices available. Therefore, during any site investigation, the Project Manager should try to anticipate the remedial options and modify the investigative scope in light of these choices.

Remedial implications influence a site investigation in two different ways:

A. Acquisition of pertinent site data needed for remedial design.

A site investigation is usually intended to collect data on the nature of contaminants, and the type of remediation or the data requirements for remedial design are very far from the investigator's mind. However, by the time the analytical testing is completed and the remedial options have been studied, the investigation has long been completed. A supplemental investigation may have to be performed, or very conservative (safe) assumptions may have to be made for the remedial design parameters, making the remediation more expensive. Many of the remedial options have unique data needs, and everything from the sample type, sample size and the sample care may be different. If these requirements were not known at the time of the investigation, it would be almost impossible to have obtained the right data. Here are three examples of remedial methods and specialized data that would be needed for design.

Remedial Option	Pertinent Data
1. Vapor Extraction	Air Permeability of core samples
2. In-Situ Bioventing	Bacterial Count (cfu's), and nutrient level
3. Excavation	Shear Soil Strength

For vapor extraction design, it is necessary to know the air permeability of soils in the contaminated zone. It is important to obtain relatively undisturbed cores for this purpose so the soil fabric and the natural grain arrangement is preserved. Tube samplers with inner liners or rings, pushed or driven with a drop hammer; provide appropriate samples, although the core diameter requirement should be checked. In situ bioventing will require knowledge of the bacterial concentrations in the site soils. This parameter, often referred to as the bacterial count, can be determined by any agricultural laboratory. However, the samples should be protected from extreme temperatures, and delivered to the laboratory within a day or two after being obtained. The shear strength testing requirements are similar to those described above for air permeability measurements.

B. Incorporation of pilot or field scale testing into the investigation.

Stand-alone pilot tests are an expensive proposition and may be considered to be a luxury. Sometimes they are performed just before remediation begins, almost as an afterthought and may not provide much information on the feasibility of the remedial method. It is possible to perform some limited pilot testing during the investigation, and the investigative equipment may have to be adapted for the pilot test. Here are three examples.

Remedial Option	Pilot Test During Investigation
1. Pump and Treat	Aquifer testing in monitoring wells
2. Excavation	Test pit to assess side wall stability
3. Ex-Situ Bioremediation	Benchscale study of bacterial effectiveness

Although a full scale pump test requires a large diameter well and pumping equipment that is not available during the investigative field work, some limited information on the hydraulic properties of an aquifer may be obtained by doing some testing in monitoring wells. If pumping is not feasible, at least slug testing may be done to make measurements useful in deciding whether a pump and treat system can be operated conveniently. If investigative trenching is being performed during the investigation and excavation is one of the options being considered for remediation, a test pit should be done to observe side wall stability for excavation planning. A bench scale study for ex-situ remediation may need a large bulk sample, and an early study may give a head start for remediation.

1.6 BUDGET, SCHEDULE, AND COMMUNICATIONS STRATEGY

For site investigation and remediation projects, the budget, schedule and the communication strategy are as important as the technical objectives. The project tasks may be first defined in terms of budget and controls, and technical details filled in later. It is essential that a project manager overseeing the technical objectives of the investigation or remediation has some grasp of project control fundamentals—even if it is with the help of a specialist assistant; otherwise, it is quite likely that the technical manager will be overruled by someone who controls these items. For an investigation or remediation project lasting more than a couple of months, a portion of the budget should be set aside to track, hindcast, and forecast the expense and schedule details.

The actual scope of this task will depend upon the complexity of the project, and the number of clients, project team members, and subcontractors. The project control budget can range from as little as 2% to about 10% of the overall project budget. Project deadlines have a way of creeping up on project managers; it is best to set aside the controls budget and use it for some of the following items:

- Project tracking using Primavera™, Surtrak™ or other similar software
- Progress meetings to discuss budget and schedule
- Progress reporting
- Resource loading
- Task loading, and
- Cash flow Analysis

Communication strategy for a project depends upon the audience. Reports meant for review by other scientific and technical peers have to be in-depth and are fairly voluminous. If the results are to be presented at a public hearing, then simple clear charts, diagrams, and displays may be necessary. Presentations to nonenvironmental agencies such as building departments and attorneys have to be prepared for scientifically lay, but otherwise very precise, audiences. In each case, the project manager should set aside a budget, allow time, and pay attention to the communications strategy. Otherwise, his masterpiece may be incomprehensible at best, and misunderstood at worst.

CHAPTER 2

ENVIRONMENTAL SETTING

2.1 THE EARTH

The earth is about 13,000 km in diameter at the equator. A molten metal core forms the innermost region that may be responsible for the magnetic field and the rotational stability of the planet. The continents, composed of lighter rock, collide along tectonic plate boundaries to form mountains and provide vents that develop into volcanoes. The seafloor, composed of heavier, basaltic rock, is also divided along plate boundaries. Movements of plates due to seafloor spreading cause some plates to ride over others; forming subduction zones where solid crust is pushed into the mantle for remelting.

Cooling of molten magma after intrusion into the crust forms igneous rocks. The mineral proportions, crystal sizes, and arrangements are the basis for the distinct character, names, and properties of various igneous rock types. The igneous rocks include basalts and granites, and are some of the hardest and most massive rocks. Igneous rocks provide many of the massive deposits where waste disposal and isolation is being considered. Folding, fracturing and faulting are as important in providing contaminant transport pathways as the rock porosity and permeability.

Over time, tectonic processes cause folding, faulting, and fracturing of the rocks. Weathering agents cause their breakdown into fragments, which are then transported and deposited by water, air, and glaciers. Over geologic time, the deposits of rock fragments are compressed into sedimentary rocks. Sedimentary rocks, with names such as sandstone, marble, and shale result from different depositional processes and derive their character and physical properties from their mineral composition, particle size, grain arrangement, age, and depositional history.

Metamorphism is the process of transforming igneous or sedimentary rocks by the action of heat, pressure, and chemicals. The transformed rocks are folded and fractured, are the least common, and are called metamorphic rocks. The three rock types—igneous, sedimentary, and metamorphic—comprise much of the earth's crust.

2.2 NATURE OF SOILS

Origin

Weathering agents break down rocks of all compositions into fragments, which are then transported and deposited in layers. Rock and mineral fragments that have not undergone tectonic pressure or temperature are termed unconsolidated sediments or soil. In addition to the inorganic rock and mineral fragments, some organic matter also becomes incorporated into soil due to the growth and decay of vegetation in the shallow subsurface. Near the ground surface soil often develops "horizons" or zones that have distinct characteristics due to the influence of weathering. The horizons are important for plant growth, but are too shallow to play a significant role in the spread and accumulation of contaminants.

The most distinct characteristics by which soils can be differentiated is the depositional origin. Alluvial soils, transported and deposited by the action of streams, are some of the most and widespread materials. The primary influence of alluvial action is to grind and polish the soil grains so that they become smaller and develop an increasingly rounded character as they are transported along stream beds. The length of time for which particles are transported and the vigor of the stream action shows in the degree of roundness of soil particles.

Mountain streams carry loads of particles of fairly large sizes, while in deltaic regions, stream action is limited to moving fine grained silt and clay particles. Similarly, during periods of high flow, larger amounts of sediments are carried along and deposited in thick beds, whereas the intervening periods of lower flow or dry periods result in very little deposition. The seasonal character of water flow in most streams results in alternating beds that become sorted by grain size with seasonal changes or stream shifting. This bedding or stratigraphy gives a definite striation to thick sediment sections. This layering also provides convenient flow paths for preferential flow of liquids and fluids (Figure 2-1). Recognition that sediments at a site are of alluvial origin automatically gives one the foreknowledge that fluid contaminants may have flowed along strata boundaries in a horizontal direction and, consequently, could be extracted or moved during remediation along the same boundaries. In coastal regions, shallow marine sediments are one likely location for leaked contaminants.

In temperate climates, aside from stream action, the other predominant agent for sediment breakdown and transport is glacial transport. Glacial deposits often result in very angular particles that have been literally torn out of bedrock.

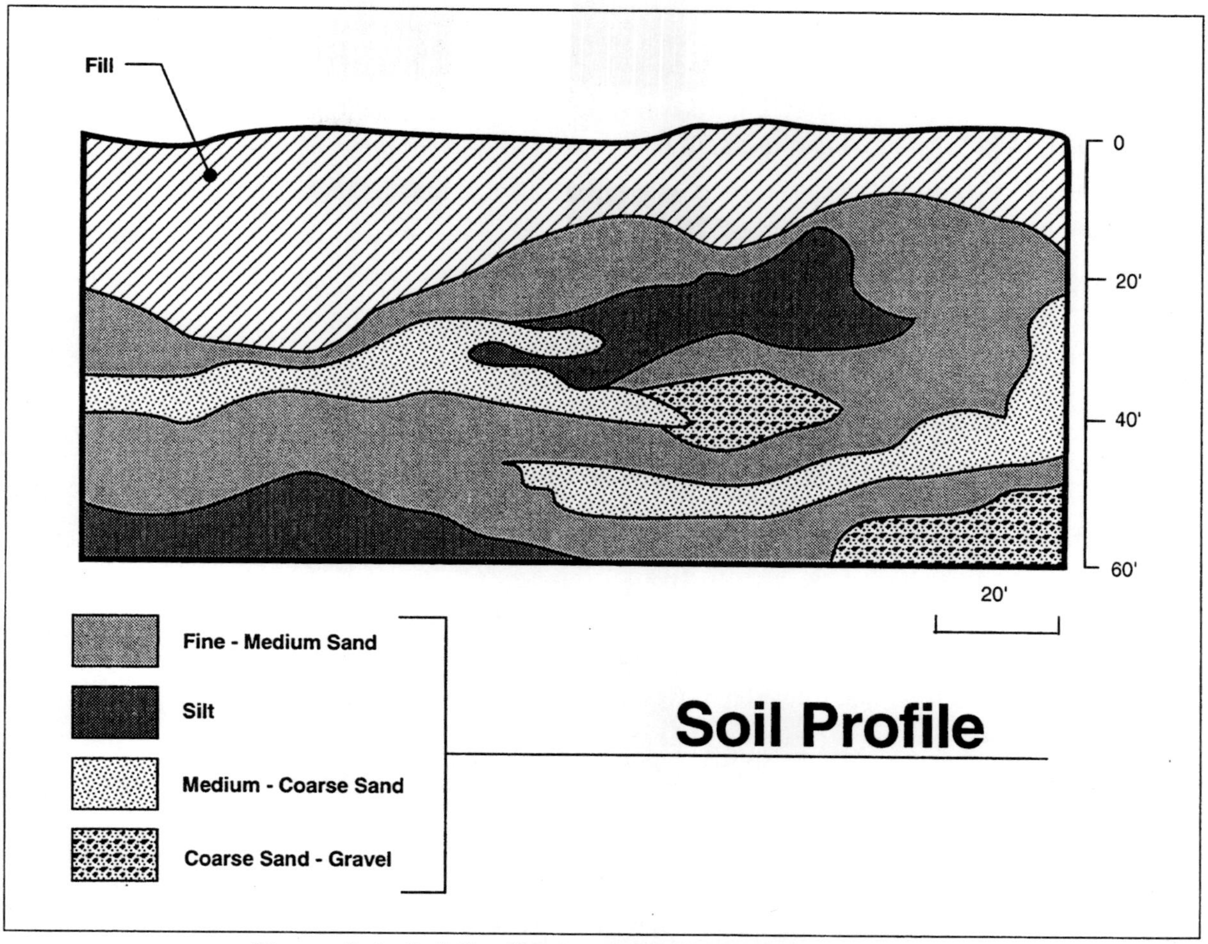

Figure 2-1. Soil Profile: Layered Alluvial Sediments

In geologic terms, the glacial ice shows a significant amount of mobility and ability to transport soils and soil particles over large distances in impressive quantities. Often, these sediments are deposited in a sudden manner in unsorted heaps of various particle sizes and arrangements. In these deposits, much of the preloading or compression may have been provided by the weight of several miles thick glacial ice on top of the sediments. The unsorted depositional mechanism leads to very characteristic deposits of till or boulder clay in which fine, cohesive clay matrix engulfs cobbles and boulder-size particles, all deposited by a retreating glacier. The till deposits of massive clay are resistant to fluid flow, but are peppered with cobbles and boulders that could reduce the feasibility of any remedial technique dependent upon drilling and installation of extraction wells.

Arid regions have the characteristic aeolian deposits that have been transported by the action of wind. The wind preferentially picks up only fine silt-sized particles and deposits them in massive sediment banks that may have very little layering and be composed of single-size particles. As water reaches these areas, sediments with cardhouse fabrics may be collapsible.

Soil Classification Systems

Identification, description, and classification of soils can be done using a number of different "systems"–sets of terms, conventions and definitions that are applicable to the particular uses for which each system was developed. Humanity has been involved in farming for a long time, and the agricultural discipline offers the most elaborate and oldest soil classification systems that have been developed by agronomists, based on principles of pedology. The soil categories in agricultural classification systems are based on the development and character of the shallow soil horizons and particle size. Soil Conservation Service and other agencies engaged in the development and promotion of farming have mapped the land surface in all fifty states using agricultural soil classification systems. These maps can provide limited clues about the character of the shallow material at a site.

Another area of endeavor where soils have been used is earthwork construction. Requirements for material handling, processing, and construction using earth materials led to the need for soil classification. These classification systems, of which the most well-known is the Unified Soil Classification System, depend on the behavior of soils when they are excavated, mixed with water, compacted or otherwise processed for construction of embankments and roads. Certain definitions such as plasticity of soils were borrowed from earlier work done for agriculture, and other indices such as gradational analysis, are similar to those used for defining aggregates for concrete.

The grain size and the mineral character of soil are the two governing characteristics that determine its behavior for earthwork construction. Based on size, soil grains can be divided into coarse-grained (cobbles, gravel, and sand) and fine-grained (silt and clay). In Unified Soil Classification System the boundary between these two main groups is the 74 micron particle diameter or the No. 200 sieve (Figure 2-2). Materials finer than the No. 200 sieve are termed fine grained and are categorized primarily on their plasticity, which is an indicator of their behavior when mixed with water. The behavior of the materials coarser than the No. 200 sieve is primarily governed by particle diameter.

Although a number of classification systems are available, the Unified Soil Classification System has gained popularity for describing the results of investigations at contaminated sites. This system is intended for excavated materials and does not include terms about the detailed grain arrangement or microfabric of soil materials. When used for describing materials at a contaminated site, descriptions related to the grain arrangement or fabric, and geologic origin, need to be added. These additional descriptions are available from geomorphology and soil science, the disciplines for study of landforms and soil formation by agents of weathering, transportation, deposition, and vegetation.

UNIFIED SOIL CLASSIFICATION SYSTEM				
SOIL TYPE		GRAPHIC SYMBOL	LETTER SYMBOL	TYPICAL DESCRIPTIONS
			AF	Artificial Fill
COARSE GRAINED SOILS				
Gravels	Clean Gravels		GW	Well-graded Gravels, Gravel-sand Mixtures, Little Or No Fines
			GP	Poorly-graded Gravels, Gravel-sand Mixtures, Little Or No Fines
	Gravels With Fines		GM	Silty Gravels, Gravel-sand-silt Mixtures
			GC	Clayey Gravels, Gravel-sand-clay Mixtures
Sands	Clean Sands		SW	Well-graded Sands, Gravelly Sands, Little Or No Fines
			SP	Poorly-graded Sands, Gravelly Sands, Little Or No Fines
	Sands With Fines		SM	Silty Sands, Sand-silt Mixtures
			SC	Clayey Sands, Sand-clay Mixtures
FINE GRAINED SOILS				
Silts And Clays (Low Plasticity)			ML	Inorganic Silts And Very Fine Sands, Rock Flour, Silty Or Clayey Fine Sands Or Clayey Silts With Slight Plasticity
			CL	Inorganic Clays Of Low To Medium Plasticity, Gravelly Clays, Sandy Clays, Silty Calys, Lean Clays
			OL	Organic Silts And Organic Silty Clays Of Low Plasticity
Silts And Clays (High Plasticity)			MH	Inorganic Silts, Micaceous Or Diatomaceous Fine Sand Or Silty Soils
			CH	Inorganic Clays Of High Plasticity, Fat Clays
			OH	Organic Clays Of Medium To High Plasticity, Organic Silts
HIGHLY ORGANIC SOILS				
			PT	Peat, Humus, Swamp Soils With High Organic Contents

Figure 2-2. Unified Soil Classification System

2.3 GEOLOGIC SETTING

Geologic setting refers to the local geology surrounding and underlying a site. A description of geologic setting should include a discussion of the broad geologic region in which the site is located. It should then describe the local geologic structure and some of the major structural components of the crustal formation in the surroundings. The description should include enough details to be able to interpret the depositional environment in which the shallow sediments may have formed. This would give an idea of the origin of soils and provide guidance regarding what physical parameters may be important for remediation. The geologic setting may provide a lot of guidance on how deep the investigative efforts need to be. For instance, in areas where no usable aquifers are known to exist, there may be no need to reach the water table.

Often the geologic setting description is a theoretical discussion that may give little clue to the behavior of contaminants that may be found at that site. However, it's important to discuss the geologic setting, to rule out any overwhelming geologic factors that may affect site investigation methodology or influence the choice of remediation methods.

2.4 LOCAL STRATIGRAPHY

Although the geologic setting provides the frame of reference for soil conditions at a site, the detailed picture is understood only when the local stratigraphy is determined. The thickness of soil and rock layers, grain size variations, density, porosity, and moisture content are the types of parameters obtained from drilling and sampling a soil boring. The variations of stratigraphy across the site are determined by drilling a number of soil borings and interpolating data between borings using the knowledge of geologic setting, as well as nonintrusive techniques such as geophysical methods.

An understanding of the local stratigraphy provides us guidance on how contaminants may have leaked, flowed, or spread in the subsurface. A cohesive zone within an otherwise homogeneous, granular cohesionless soil mass at a site can significantly alter the flow of fluids in the subsurface. A leak from an underground storage tank may trickle vertically down through coarse sand and may then start spreading laterally due to the presence of a silt zone within the coarse sand.

The amount of effort and time that is spent on establishing the local stratigraphy is proportionate to the complexity of layering as well as to the use for which this information is being sought. If the stratigraphy of the site is very simple and the soil conditions are limited to one or two major layers with horizontal strata, a small number of borings is sufficient to establish the soil conditions. On the other hand, in extremely complex stratigraphy greater effort is needed. At times, the amount of drilling and sampling may be aimed at a specific strata boundary or layer type. The layer may contain the contaminants in some preferential manner, and therefore could be a potential target for remedial efforts.

How this local stratigraphy is represented and displayed is becoming very important, as new and powerful visualization and graphics tools are becoming available. It is possible with a little effort to be able to place the stratigraphy in an exact relationship with existing facilities, utilities and any obstructions that might be in the way. The results of any past investigations at the site or in the vicinity should be checked and correlated against the known local stratigraphy at the site.

Also, any conjectures or hypotheses about contaminant transport as influenced by the local stratigraphy should be checked against known information about adjacent sites. It is possible that subtle variations in silt content may hinder contaminant transport in an aquifer while on the other hand, massive cohesive layers may have enough fractures to be ineffective as aquitards. The local stratigraphy is displayed on a boring log that shows the vertical variation and the layer boundaries between significant zones of sediments and soils. Physical information displayed includes grain size, penetration resistance (sampler blow counts), and moisture content. Results of contaminants tested are also shown. One or more cross-sections are prepared and a discussion of the predominant layering system across the site is included in the report text. If a remediation scheme that is dependent on stratigraphy has to be designed and employed, then fairly sophisticated three-dimensional representation of layering at a site may be required in order to evaluate the proposed schemes.

2.5 GROUNDWATER

Groundwater is hidden from view in subsurface layers of soil and rock (Figure 2-3), making it difficult to understand fully. The general principles governing the occurrence, movement, and quality are well understood, but all water data show variations with time (Figure 2-4). Groundwater hydrology is not an exact science by any means.

Water under the ground surface occurs in two main zones: an upper unsaturated (vadose) zone and a lower saturated zone (Figure 2-3). The boundary between the two zones is the water table. In the unsaturated zone most of the soil pores are filled with air, and water occurs in discontinuous zones adsorbed to the soil grains. A zone of capillary rise immediately above the

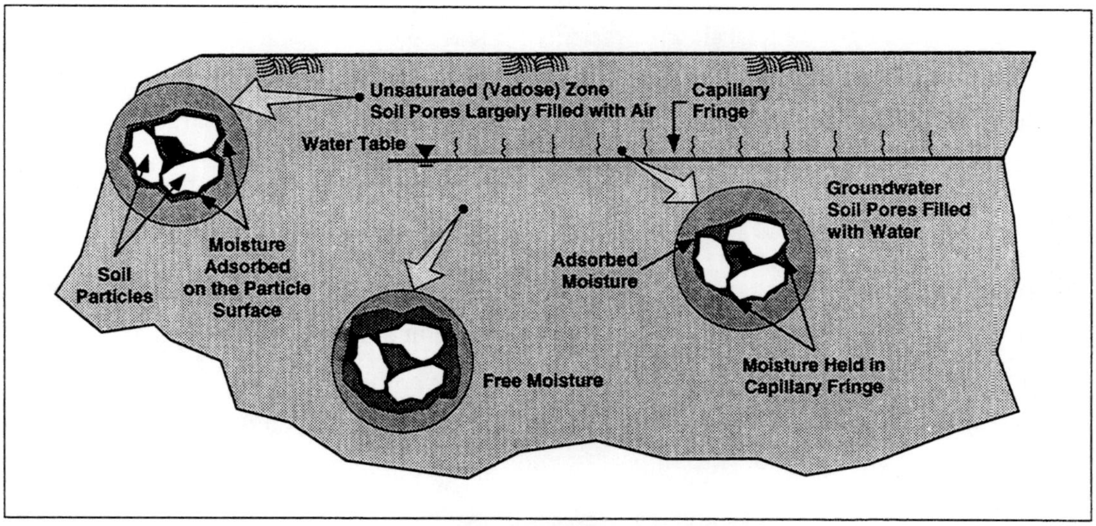

Figure 2-3 Water in Soil

water table is termed the capillary fringe. Free groundwater occurs below the water table where all the soil pores are filled with water. The elevation and configuration of the water table must be recorded with care. In clayey deposits the first occurrence of water may be several feet below the water table, and water level may rise in a boring that is left open with a standpipe. Even with accurate measurements, the conclusions drawn from different sets of wells could produce different results. Seasonal influences on the water table can cause differences of several feet between fall and spring. Construction dewatering at adjacent sites or other nearby pumping can temporarily depress the water table.

Presence of groundwater at a site within the zone of contamination is probably the biggest single complicating factor for site investigations. The contaminants become much more mobile when they are water-borne. Contaminants that are volatile or highly mobile behave differently depending upon whether they are miscible or immiscible with water, as well as whether they are denser or lighter than water. Thus, nonaqueous liquids become stratified into lighter fluids that float on water such as gasoline, and heavier than water fluids that sink down such as some of the solvent components. Thus, the spread of contaminants becomes much more enlarged as compared to a site where water table is not within the contaminated zone. Soluble inorganic contaminants such as nitrates become widely dispersed upon reaching the water table.

An investigation of static media such as soil and rock can be done within a defined interval, but the investigation of a dynamic medium—groundwater—requires observations over extended periods. Observations of contaminant concentration in an individual well will show some variations if performed over several calendar quarters (Figure 2-4). The observations record the influence of seasonal and other variations as well as the behavior of groundwater in response to some induced changes such as slug tests, pumping tests, or installation of other nearby wells. Any remediation of a site where contaminants have entered the water table also becomes much more problematic. It may only be possible to extract the contaminants a little at a time as the chemicals may have become adsorbed to the soil grains.

Pumping and treating of groundwater only allows limited flushing action of the aquifers where contaminants might be located. Sometimes a single point source cannot be positively confirmed as the only location from which contaminants may have originated, and there is a possibility of multiple sources. In such cases, site-by-site remediation of contaminants in the groundwater becomes futile, and regional cleanups extending over whole aquifers have to be devised and carried out.

An understanding of the contaminant behavior and the potential effectiveness of remediation can often be gained by performing some computer modeling of the contaminated groundwater flow in response to the remedial action. In addition, the results of modeling may have to be displayed with the aid of graphics and visualization software to be able to make rational decisions about investing in remedial efforts.

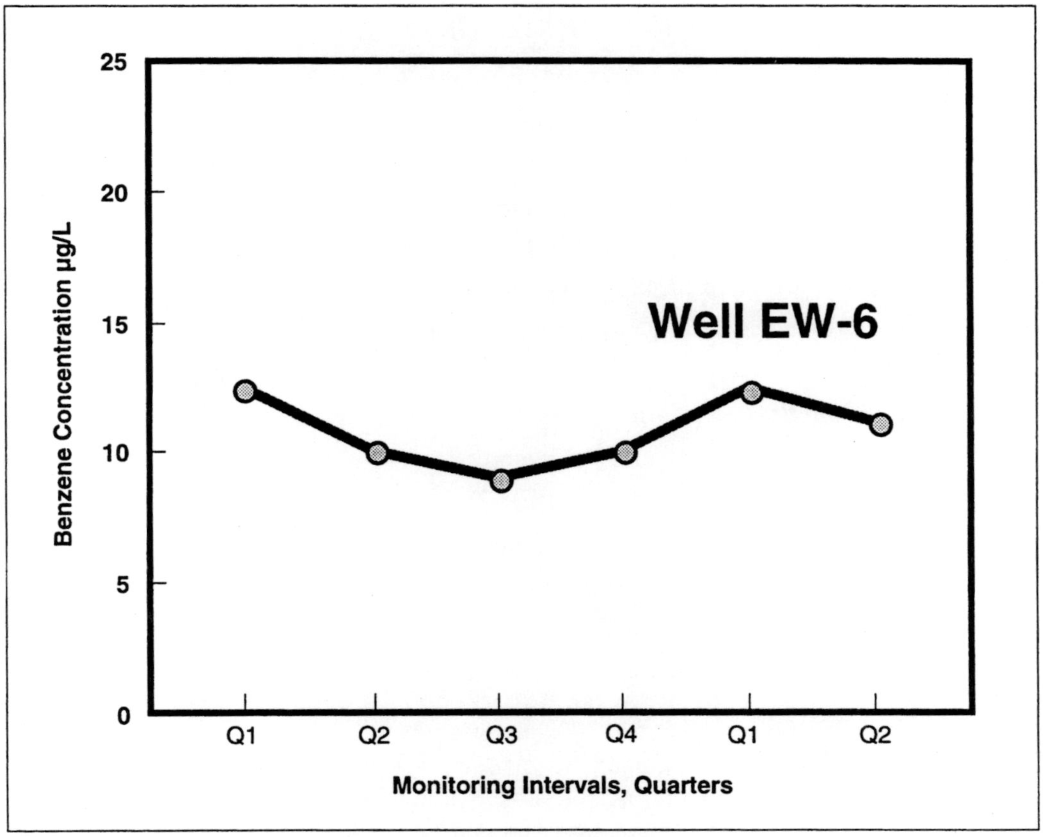

Figure 2-4. Well EW-6

2.6 CHOICE OF INVESTIGATIVE METHODS

Environmental setting of a site can reduce or increase the effectiveness of a particular investigative technique. Even though drilling, sampling, and laboratory testing are necessary at every site to obtain data, the choice of drilling techniques, the depth to which investigations are to be performed and the spacing of borings are influenced by the environmental setting. Success of geophysical methods, resistivity techniques, and soil gas surveys is also dependent on the soil types. Massive clayey soils with very low permeability may not yield usable information during soil gas surveys and the presence of large amounts of salts may render the soils unsuitable for use of resistivity techniques.

Local practice governs the choice of drilling, sampling and testing equipment available in any particular region. For instance, in glacial environments, drilling equipment has to be able to break, penetrate, and drill beyond any large cobbles or boulders. In areas where borehole collapse and presence of groundwater are prevalent, hollow stem auger techniques become unfeasible and rotary drilling has to be employed. Review of nearby drilling records and a limited amount of geophysical data acquisition is sufficient to decide the effectiveness of any particular technique.

2.7 LIMITATIONS OF REMEDIATION TECHNIQUES

Even though remediation is aimed at contaminating chemicals, and soil, rock, and groundwater are considered inert media, they become significant in selecting remedial techniques. During remediation, very small quantities of contaminants are handled, but large amounts of soil, rock, and groundwater need to be processed before the contaminants can be accessed and separated. If in-situ methods are planned, the presence of massive clay deposits may make vacuum extraction unfeasible. Presence of a few granular, high porosity layers governs the length and placement of well screens in any type of subsurface extraction technique. For ex-situ methods, often the handling and processing of soil and rock become the main physical effort in remediating a site. Soil washing, thermal desorption, or even dig-and-haul offsite disposal all are dependent on the soil and rock type present. The choice of excavation equipment, handling equipment, sieves, shakers, and other soil handling techniques are all influenced by the grain-size and soil type. If the treatment equipment is only capable of handling a certain particle size matrix proportion, then crushing, separating, and other pretreatment processes may be required before effective remediation could be performed.

While nearly all remediation techniques may be possible in all environmental settings, certain ones may prove to be much more effective, while others may not be feasible. The environmental setting of a site may rule out certain remediation techniques, or it may make other techniques more effective.

CHAPTER 3

INVESTIGATIVE TECHNIQUES

A number of techniques have been developed for investigating the nature and distribution of contaminants at a site. The investigations are generally conducted in two phases: Phase I Environmental Site Assessment (ESA) is done without sampling, while Phase II Investigation or Site Characterization includes intrusive sampling and testing.

There are three main components of a Phase I investigation, site inspection and interviews, historical document search, and review of agency lists. These are described in Sections 3.1 through 3.3. Depending upon the background of each site, one of the three techniques may take on greater importance. Some very detailed investigations are considered Phase I studies, as long as they do not involve intrusive sampling.

Remote sensing and nonintrusive geophysical techniques may be used either during Phase I as a final check of suspected subsurface features, or may form the preliminary step used to plan the intrusive sampling and testing. These are described in Section 3.4. Phase II investigations use intrusive techniques and are described under three main headings: Probing Techniques (Section 3.5), Drilling and Sampling (Section 3.6), and Groundwater Investigations (Section 3.7). Analytical Methods for testing of contaminants are surveyed in Section 3.8.

3.1 SITE INSPECTION AND INTERVIEWS

An initial visit to a site suspected of being contaminated is crucial, and the purpose is three fold:

1. Observing the *current conditions*, which may include ongoing industrial activities for a facility that is still in operation;

2. Forming an impression about the site's *past operations*; this would be of primary importance for an abandoned brownfield facility, but is also significant for noting the changes that may have taken place in a currently operating facility.

3. *Selecting investigative methodologies* by looking at the opportunities and limitations that the site offers, and what intrusive methods are feasible at the site.

A number of good checklists are available for recording the *current conditions* of a property, and these should be used to advantage. A suggested checklist is included in the Appendix. Less guidance is available on how to accomplish a review of past activities or to plan site access for intrusive investigations. These two items are addressed below:

The best way to learn about *past operations* of an abandoned brownfield facility is to obtain a site map from the era when the facility was in operation and tour the premises accompanied by someone familiar with past operations. If only a map is available but no knowledgeable individual can be found, a lot of the interpretation is left to conjecture. A former employee in the absence of any documents is also of limited assistance, as he or she has to rely exclusively on memory. If more than one site layout document is available from the era—a plot plan and an air photo—then highest reliance should be placed on items that can be identified on both.

The primary purpose of a historical review is to identify items that require direct, intrusive investigation. Because of time lapse, it is easy to miss something significant, a closed dump, a large sump, or the location of some long-term UST leak. The significance of objects on a plan may be known to someone knowledgeable about practices from that era even if this individual was never connected with the particular facility under review.

The third main objective of a site visit is to *select investigative methodologies*, including intrusive techniques, if and when required. Both a general or site-wide strategy, and techniques aimed at investigating localized contamination should be selected. The overall strategy should recognize the physical limitations of the site, such as presence of buildings, in which all drilling and sampling may need to be done using limited access rigs. A long linear site such as a railroad track or a pipeline right-of-way may require very widely spaced shallow samples.

Sampling locations are selected on the basis of both the current features (barrel storage area, clarifiers, oily spots) and the former locations of past operations (past UST locations, former paint shop, etc). If the target locations are now occupied by buildings, slant borings may need to be used.

3.2 HISTORICAL DOCUMENT SEARCH

General

In investigating a site, it is important to try and determine the past site use and its influence on the nature and distribution of any suspected contaminants. The environmental assessments have borrowed historical search techniques from a number of different fields where there has been need for searching the past site uses. Some of the techniques come from utility surveys where the past location and type of utilities are investigated in order to plan future construction or installation of additional utilities. Other techniques have been derived from title search analysis where the exact ownership is an issue in determining future sale or purchase of a property. Archival search techniques such as those used by industrial archaeologists and historians have also been adapted. Each of the historical pieces of information has to be kept in context. We have to realize that there may be a degree of uncertainty in the accuracy and completeness of any historical record. In fact, the main purpose of looking at a number of alternate and dissimilar sources is to ascertain the accuracy of some of the important and significant items.

How far back does the record need to be checked is a function of where the site is located. On the West Coast of the United States, going back to the 1920s puts us in the ancient era of industrial development. For the East Coast of the United States, often a 100-year old record is not out of question. In industrial Europe, this may need to be extended over 200 years to the earliest time of industrial revolution when availability of chemicals and disposal of industrial wastes began.

Aerial Photographs

Historical aerial photographs are direct physical references for reviewing past conditions beginning in the early 1920s when flights became common and aerial photography became available. Some of the pioneers of aerial photography have left behind valuable archives of early photographs. These are held by educational institutions or libraries and are available to anyone to look at for a small fee after an appointment. Similar archives are also available commercially in different parts of the world, extending back to the early part of the century.

When investigating an earlier era in a photograph, the interpretation of objects seen is only as good as the knowledge of the person doing the interpretation. One has to be familiar with both how objects appear in aerial photographs and also with the industrial history of the era when the photographs were taken. It is often useful for the interpreter to simply describe what is being seen in notes, and allow for some degree of uncertainty in the identification of objects so that modifications could be made based on alternate historical documents.

When photo interpretation is combined with the knowledge of objects that may still be in existence, the location of objects no longer in existence can be ascertained, calculated, and surveyed with reasonable accuracy, and allows for planning the site investigations.

Historical Maps

Historical maps often predate any industrial development at the site and may have been made for purposes of town planning, future developments, tax assessments, etc. Governmental agencies including the United States Geological Survey and other land survey organizations have prepared maps for geographic and historical interest as a record of natural resources, for assessing changes in land use and developing a tax base. In using these maps it should be realized that what may be interesting and useful to us today may not have been very important for the map maker. The degree of detail and accuracy may be fairly good for objects related to the original map use, but is often variable across a single map.

When using any of the historical maps the purpose for which the map was made should be kept in mind. One must realize that parts of the map where development had already taken place may be fairly precise, but other areas may just simply have been a visual approximation for "parts unknown." Boundaries of such objects as disposal areas, swamps, and natural lagoons may be very approximate. Also, the names given to different streets or buildings may have changed over the years. Objects in existence today such as a street may have had a different name earlier, or the earlier names may be retained although the street may have been relocated or realigned.

Channelized water courses, fences and boundaries may have been shifted around to conform to other land use and may simply not have been fully recorded on maps. Fire insurance maps such as those prepared periodically by Sanborn Fire Insurance Map Company were prepared to record the position of flammable, or combustible items. Therefore, the position or even the existence of some of the objects that did not concern the fire insurance inspectors directly has to be looked at with some degree of uncertainty. As-built building maps are very important municipal resource, but are notoriously inaccurate in portraying the exact locations of individual utility lines across a property. Comparison and conformance between maps and aerial photographs are potentially the most authoritative, but also sometimes the most frustrating step in historical archival interpretation. Objects that can be confirmed from two independent historical sources have a very high degree of certainty.

Title search information showing the ownership of the parcel is usually checked as a part of due diligence for property transfers. It's possible that benign sounding "mom-and-pop" owners may have leased the property to someone who had a potentially harmful site use such as fuel storage or disposal of industrial wastes. Municipal permits for building construction, building additions, road alignments, installation of underground tanks, etc. may contain sketches of property boundaries, and descriptions of objects being installed or demolished with an engineering accuracy. These records are available from the Building Departments, Planning Departments, or Fire Departments of the various city or county agencies that have jurisdiction over the site. It's important to remember that this jurisdiction may have shifted over the years and overlapping reference checks may need to be made.

3.3 REVIEW OF AGENCY LISTS

A number of regulatory agencies publish lists where potentially contaminated areas and properties are listed. These lists include such large and significant lists as the National Priority List which includes all properties placed on the Superfund list, indicating large industrial sites capable of polluting broad areas. Other lists published by states include lists of toxic pits, where large-scale disposal of industrial wastes has taken place in the past. Regional Water Quality Boards have published lists of all municipal or other landfills that require groundwater monitoring wells to check for leachates. Local fire departments have sometimes published lists of reported leaking underground fuel storage tanks. In the early days of Phase I site assessment studies (early to mid 1980s), list search checks were one of the most laborious and expensive tasks that had to be performed for reviewing a contaminated site. Computerized databases on which most of the listed items have been included now allow rapid electronic searches within prescribed radii around the site for a modest fee.

Although guidance is available from standards such as ASTM regarding the radius in which search should be made, the distance may need to be modified with a knowledge of the area, the site, the contaminants and the groundwater regime surrounding the site. For oblong sites, a series of contiguous radii are usually searched. The expertise required these days is not in going through all the lists, since that's done electronically, but in interpreting the results that are received. The computerized search results are often a map showing the location of all objects that appear to be significant, accompanied by a table. An individual knowledgeable about the area should be consulted to decide which items require further effort in the form of personal visits to agency offices for file review. A focused file review is often the most significant step where useful information is likely to be found that may require further follow-up. Routine inquiry letters for updates of any published or known information from the lists should be sent to all concerned agencies. Replies may not be received on all letters in time available for performing Phase I assessments.

The key federal lists reviewed for Phase I Assessments are described below.

- CERCLIS List–This list contains sites which are either proposed to or on the National Priorities List (NPL) and sites which are in the screening and assessment phase for possible inclusion on the NPL. The information on each site includes a history of all preremedial, remedial, removal, and community relations activities or events at the site, financial funding information for the events, and unrestricted enforcement activities.

- ERNS–The Emergency Response Notification System is a national database used to collect information on reported releases of oil and hazardous substances. The database contains information from spill reports made to federal authorities including the EPA, U.S. Coast Guard, National Response Center, and Department of Transportation.

- NPL–National Priorities List is the EPA's database of uncontrolled or abandoned hazardous sites identified for priority remedial actions under the Superfund program. A site must meet or surpass a predetermined hazard ranking system score, be chosen as a state's top priority site, or meet three specific criteria set jointly by the U.S. Department of Health and Human Services, and the U.S. EPA in order to become an NPL site.

- RCRA–The EPA's Resource Conservation and Recovery Act (RCRA) program identifies and tracks hazardous waste from the point of generation to the point of disposal. The RCRA Facilities database is a compilation by the EPA of facilities which report generation, storage, transportation, treatment or disposal of hazardous waste.

- Large Generators–Large generators are facilities which generate at least 1000 kg./month of nonacutely hazardous waste (or 1 kg./month of acutely hazardous waste).

- Small Generators–Small generators are facilities which generate less than 1000 kg./month of nonacutely hazardous waste.

- TSD facilities–Facilities which treat, store, or dispose of hazardous waste.

- Violators and Enforcements–RCRA violators are facilities which have been cited for RCRA violations at least once since 1980. RCRA enforcements are enforcement actions against RCRA violators.

- Corrective Actions–The EPA maintains this database of RCRA facilities which are undergoing "corrective action." A "corrective action order" is issued where there has been a release of hazardous waste or constituents into the environment from a RCRA facility. Correction actions may be required beyond the facility's boundary and can be required regardless of when the release occurred, even if it predates RCRA.

In addition to the federal lists, a number of state, county, and city lists are checked. An example of a comprehensive local list is the Cortese list in California. The California Governor's Office of Planning and Research annually publishes this listing of potential and confirmed hazardous waste sites throughout the State of California. The database for the Cortese list is based on input from the following:

1. CALSITES–Department of Toxic Substances Control, Abandoned Sites Program Information Systems;
2. SARA Title III Section III Toxic Chemicals Release Inventory;
3. Department of Toxic Substances Control, Hazardous Waste Information System;
4. State Water Resources Control Board;
5. Integrated Waste Management Control Board (solid waste facilities);
6. Air Resources Board, dischargers of greater than 25 tons of criteria pollutants to the air;
7. Air Resources Board, dischargers of greater than 10 and less than 25 tons of criteria pollutants to the air;
8. Leaking Underground Storage Tanks;
9. Underground Tanks;
10. Inventory Update Rule (chemical manufacturers);
11. Waste Board-Leaking Facility (site has known migration);
12. Waste Discharge System-Enforcement Action;
13. Department of Toxic Substance Control Docket.

3.4 NONINTRUSIVE METHODS: GEOPHYSICAL TECHNIQUES

The historical search may point to a number of suspected locations where subsurface items may have been left behind. These may include underground fuel storage tanks, sumps, clarifiers, basements, utility lines, or disposal areas that may have been covered up. Identification and confirmation of these suspected objects can sometimes be made by using nonintrusive sensing techniques collectively termed geophysical methods. Another application of these techniques is during the planning of site investigations. Locations of utilities and underground structures need to be confirmed before undertaking intrusive activities.

As the name implies, the geophysical methods depend upon the physics of the earth to try and detect any anomalies or irregularities in the subsurface. If we knock on wood and then on metal, different acoustic reflection signals come back, revealing the nature of the material on which we have made an impact with our knuckles. By knocking on a surface, we can also get an indication of whether we are knocking on a solid object or a hollow cavity.

The geophysical methods employ a source for either sending an impulse or projecting a field into the subsurface. The geophysical techniques use a number of different material properties including electric, electromagnetic, magnetic, or elastic characteristics. An electrical field is set up by inserting electrodes on a certain pattern below the ground surface and setting up an electrical current. Magnetic or electromagnetic signals, pulses or fields are projected from sources that can be moved, dragged, or towed around the site along certain preplanned grid lines to try and get a complete coverage. The receivers can be electrodes that are inserted into ground or antennae such as the EM wand. Any variation in density, elasticity, electrical, or magnetic properties will tip us off about the presence of a different material or object by distorting the field that's being projected or through a return signal that's not representative of homogeneous isotropic media.

Traditionally, experienced interpreters were needed to explain the crude and sometimes seemingly incomprehensible return signals and plots. One had to know about false reflections, shadows, duplicates, and internal quirks of the generating sources, the antenna, the receivers, the lengths of wiring and the influence of extraneous electrical or magnetic objects to make an accurate interpretation. While the need for some expert interpretation will always remain, great advances are being made in computer enhanced image processing and representation of the physical information. Many of the developments in this area have been initially made in the medical field to interpret the inner workings of the human body, or in resource development areas and oil exploration. The net result has been that the images produced and the interpretation of data is of much higher caliber and certainty. Now we need not only an experienced geologist who can make interpretations, but also individuals who are graduate level, highly trained computer engineers. Brief descriptions and applications of the major techniques are given below.

Magnetic and Electromagnetic Methods

The magnetic methods generally consist of an electromagnet or other source that establishes a field in the subsurface, and a sensor that detects any distortions in the field. These methods are primarily useful for ferrous objects such as underground steel tanks, buried metal conduits, reinforcing bars, steel beams or metal drums. For nonferrous objects, the magnetic methods are of limited use, if any. Even some ferrous objects, such as a steel tank that has rusted or deteriorated, may not be detectable by the use of magnetic methods. Their best application has been in finding underground conduits buried at shallow depths which indicate the position or presence of larger objects, for example, a filler pipe leading to a buried tank. They obviously have an application in locating active utilities such as water or gas lines or abandoned utilities such as heating oil lines.

Ground Penetrating Radar

These instruments have been very useful in producing records of the shallow subsurface for

many types of objects. The instrument works by sending a radio frequency wave into the subsurface and then receiving signals returned by reflection. The material property that primarily affects the propagation of radio frequency waves, the dielectric constant, changes across material boundaries, thus producing reflections. The penetration of radio waves is fairly limited into the subsurface so that these instruments have primarily been used for the upper 2 or 3 meters.
The source and the antenna are often mounted on a sled or a small trailer that can be towed by hand or behind a small vehicle, depending on the distances involved. A grid pattern is traced using the instrument, and a map is produced showing any reflections or anomalies that may have been detected.

Impact Techniques

Impact sources range from a sledge hammer struck on a steel plate sitting on the ground surface, to charges of explosives that are used to create a pressure wave front emanating from the source and travelling into the subsurface. This pulse or wave is reflected or refracted back to a sensor based upon changes and variations in the elastic properties of the various layers along its path. The reflected or refracted waves are then sensed by using geophones or other sensors inserted into the shallow subsurface. Knowing the travel velocities, and the distance between the source and the sensors, a reasonably accurate interpretive map of the subsurface layering can be prepared.

When using geophysical methods, the point of diminishing returns is reached fairly quickly. Spectacular and surprising results may be obtained with a small amount of effort while only modest gains in accuracy and knowledge about the subsurface may be achieved by spending a lot of additional time and money. It is common to have interference from overhead power lines, adjacent metal buildings, parked cars in the vicinity or other above-ground objects that interfere with the travel and sensing of electrical and magnetic pulses and fields projected into the subsurface.

3.5 PROBING TECHNIQUES

A number of rapid probing and sounding techniques are available to investigate the subsurface conditions at a site. Two main advantages of probing techniques are rapidity and minimal drill cuttings. These techniques range from small 1/2-inch diameter hand-driven steel tubes capable of penetrating to 2 or 3 meters in soft or moderately hard soils, to high capacity sounding equipment. CPT (Cone Penetrometer Test) probes use a 1-inch steel rod pushed hydraulically by a motorized truck-mounted equipment capable of penetrating to 100 meters below the ground surface. At the desired target depths, fluids can be withdrawn using suction or motorized pumps to obtain samples of soil gas or groundwater. In simple soil gas or soil vapor surveys, syringes may be used to transfer samples of the soil vapors for testing in a mobile laboratory stationed at the site. The laboratory has equipment capable of analyzing the presence of volatiles that give an indication of the presence of contaminants in the soil or groundwater. The CPT type equipment is also used with sampling tools that can obtain small samples of soil or groundwater for analysis. Other intermediate capability equipment uses pickup trucks as a reaction and is capable of penetrating down to about 10 meters below the ground surface.

Three examples of the equipment types and their application are given below:

1. *Hand-Operated Soil Gas Survey Equipment*

   This equipment is primarily built around a mobile lab capable of rapid analysis. It includes a laboratory van housing testing equipment for volatile organics and fuels. In addition, it uses a hand drill capable of making a 1/2-inch diameter hole up to 20 feet below the ground surface. A 1/2-inch steel diameter tube containing Teflon tubing is inserted into the hole. The air sample is withdrawn using syringes or a small vacuum pump and then tested in the GC-MS (Gas Chromatograph-Mass Spectrometer) equipment for analysis.

2. *Geoprobe™*

   This is a narrow diameter, direct push method in which a soil sampler attached to a steel rod is pushed down to the prescribed depth by using a vibratory head. The vibratory head includes a hydraulically operated hammer assembly that is attached to the 1½-inch diameter steel rod. The vibratory hammer is operated by using hydraulic power from the drivetrain of a 4-wheel drive pickup truck and along with sampling tools is transported in the truck. The compact arrangement makes the overall package very mobile. A small amount of cuttings is produced, and the hole can be plugged using bentonite and topped off with an asphalt or cement concrete patch.

   A tube sampler with a liner is used to obtain soil samples. The soil sample obtained is 1.06 inches in diameter and 22 inches long, in a plastic or brass liner. For groundwater sampling, a 22-inch long screened sleeve is installed and water is allowed to enter the screen. If desired, the temporary well can be "developed" in preparation for water sampling. Water is withdrawn using a suction pump and a narrow diameter hose, and collected in sampling containers at the ground surface.

3. *CPT Equipment*

This equipment is truck-mounted, hydraulically operated, and pushes against a reaction of several tons with a force of several thousand pounds. It pushes a 1-inch steel diameter rod with a cone attachment. The attachment can be changed and soil vapors, soil samples, or groundwater can be obtained for analysis.

3.6 DRILLING AND SAMPLING

General

Drilling and sampling of soil and rock provide the most definitive method of obtaining a sample from a specific depth for testing and analysis. Many of the action levels, cleanup criteria, and regulatory requirements governing the disposition and fate of a site are based on analytical test results performed on individual soil samples obtained by conventional drilling and sampling. The techniques used in such work are fairly routine and a large number of experienced drillers and technicians are available on a competitive basis. The protocols for sampling and recording information are well established, and extensive experience has been gathered in collecting and interpreting such data for a number of purposes. Historical uses of these techniques have included agricultural surveys, geologic explorations, mining explorations, and geotechnical engineering and foundation investigations. Interpolation between borings requires experience and an understanding of geology. Such experience and expertise is possessed by a number of moderately to highly experienced geologists, civil engineers and environmental professionals.

Spacing of Borings

In order to select the spacing of borings, some information must be available about the character of the subsurface materials and the source of contamination. An investigation should begin with a few widely spaced borings, and if additional data are needed intermediate borings should be added. Spacing should be smaller in areas of greater soil variability or higher concentration of contamination.

Hand Augering

A hand-operated soil auger is the simplest equipment for making shallow holes in the ground and securing cutting samples of the soil material in a loosened condition. Augers can be used to drill holes to depths of 10 to 15 feet in diameters from 2 inches to 6 inches. Small earth augers are fitted with handles, and extensions can be added to the handles.

Augers have the advantage of rapid drilling, but the depth from which the drill cuttings originated can only be approximated. Auger cuttings provide a disturbed sample that is useful for determining the soil classification and moisture content, but undisturbed samples must be obtained by driving a sampler down the augered hole.

Soil Borings

Soil sampling is done in two steps: drilling of a soil boring to reach the depth of interest, and pushing or driving a sampler to obtain a representative and relatively undisturbed sample. Drilling is usually done by using a motorized drill rig, fitted either with an auger (solid-stem or hollow-stem) or, a high speed cutting head and circulating water (rotary drilling) (Figure 3-1). In soft clays or sands, drilling mud is required with rotary drilling to keep the hole open. All drilling tools are steam cleaned before drilling and between borings.

The most commonly used sampler, called a split spoon or split barrel sampler, consists of a steel tube split lengthwise (Figure 3-2). To the lower end is attached a cutting shoe, to the upper, a check valve and a connector to the drill rods. The size used for Standard Penetration Tests (SPT) is 1.45-inch ID and 2-inch OD. Other larger samplers with 2-inch IDx2.5-inch OD and 2.5-inch ID with 3-inch OD are also used. These larger diameter samplers often include rings, sleeves or liner segments to retain the samples.

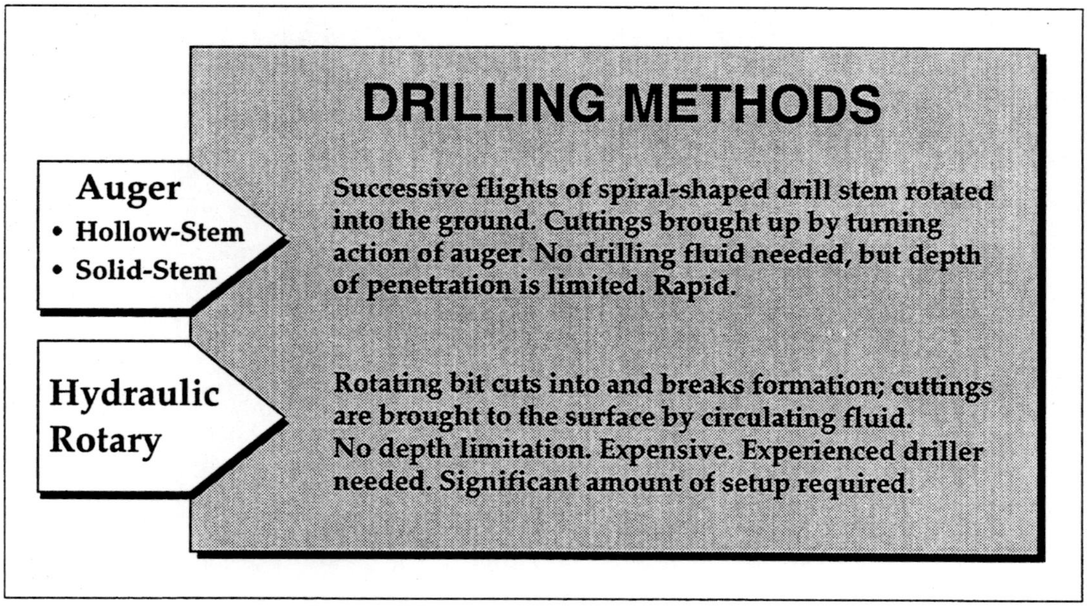

Figure 3-1. Drilling Methods

Samplers are washed thoroughly in solutions of TriSodium Phosphate and then rinsed in deionized water. Sampling is done at nominal intervals of about 5 feet; but the sampling interval could be half of that in a heterogeneous material at shallow depths and twice or more at greater depths in uniform soils. In addition, samples are taken at all changes in stratigraphy. At sampling depth, drill tools are removed and the sampler lowered to the bottom of the hole by attaching it to the drill rods. It is then driven 18 inches in 6-inch increment with a hammer. In Standard Penetration Tests, the hammer weight is 140 pound and its drop is 30 inches. The number of blows for each 6-inch penetration is recorded. The Standard Penetration Resistance, or N-Value, is the sum of the number of blows for the second and third 6-inch increments.

The penetration resistance, or N-Value of a soil is a measure of the density of sands or consistency of clays. Even if a standard sampler or a standard hammer is not used, blow counts are recorded and reported as blows per foot. The hammer weight and drop and the Sampler ID and OD must be recorded in the field so the blow counts could be converted to equivalent N-Values for comparison and for evaluating the consistency of soil deposits.

After bringing the sampler to the surface, the soil is examined and classified by a field engineer, geologist, or technician. The examination includes looking at the soil in the cutting shoe. One ring or short liner tube may be extruded in the field for examination and classification. Organic Vapor Analyzer (OVA) or other similar instruments are used to monitor vapors from the borehole by inserting the instrument's sensor tip a few inches into the borehole. A measurement should also be attempted after splitting an extruded sample. Sometimes the soil used in classification is placed in a small polyethylene Ziploc bag, loosened and then monitored for organic vapors by inserting a sensor tip into the bag. The level of organic vapors, expressed in parts per million, and how the measurement was made, should be recorded on the boring log.

The soil or rock sample is then packaged and preserved for transportation to the laboratory. The tube ends are covered with Teflon strips and closed with plastic end caps. A label is affixed to one end with the following information: date, project number, owner, site, boring number, sample number, depth, and name of the person recording the information. The sample is kept cool in an ice chest with some blue ice to maintain its temperature at 4°C until arrival at the laboratory. A chain of custody form is completed at the site and shipped with the samples.

3.7 GROUNDWATER INVESTIGATIONS

Groundwater Contamination

The quality of groundwater can vary over a broad range, and it is often difficult to distinguish the origin of water-quality problems. Natural quality of groundwater is affected by the type of mineral substances with which it has come in contact. For centuries, man has disposed of waste by throwing it in streams, storing it above the ground, or burying it below ground. The quality of groundwater is strongly affected by waste disposal and land use. Here are some examples of groundwater quality problems. These problems can originate on the land surface, above the water table or below the water table.

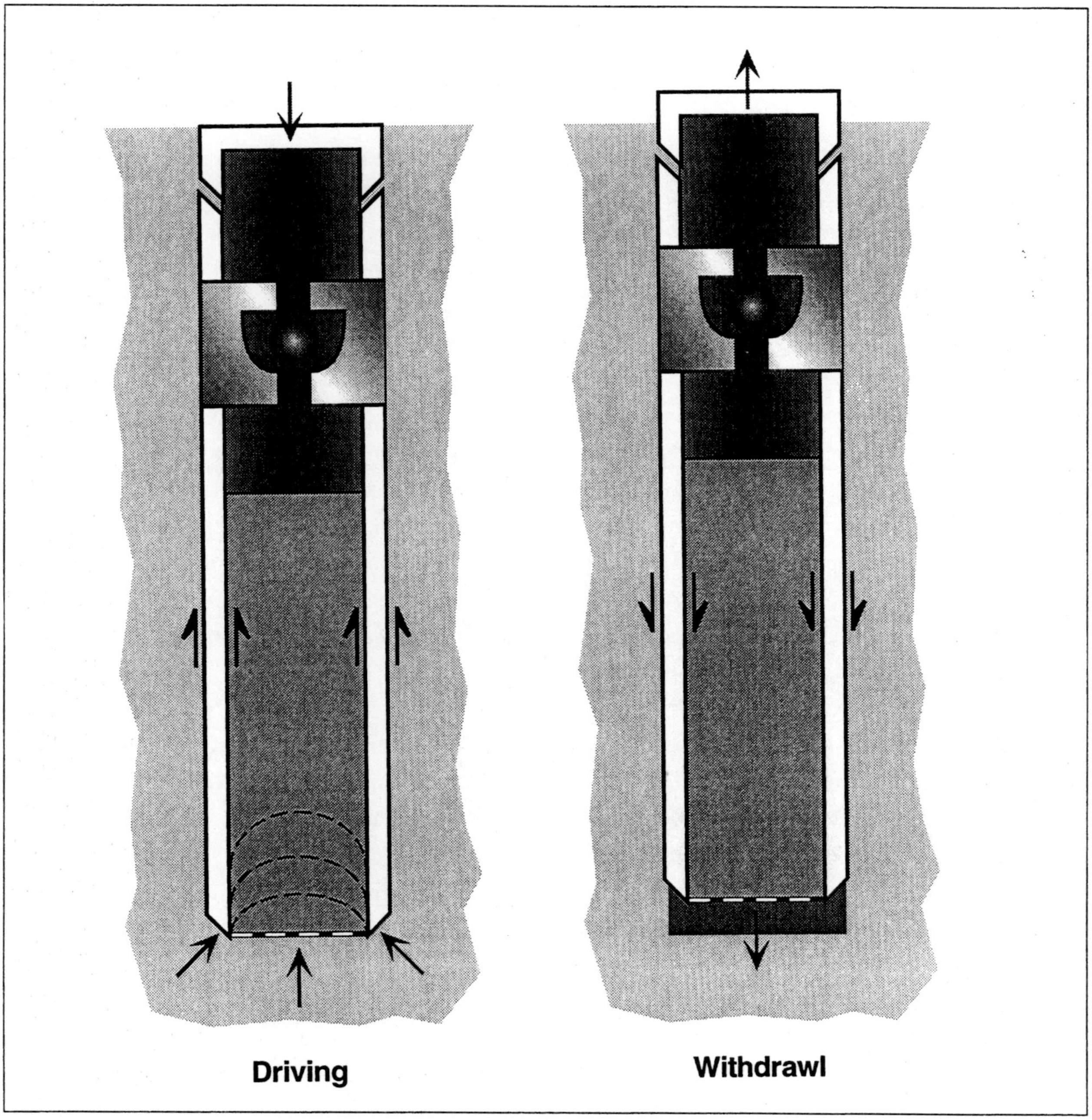

Figure 3-2. Forces on Sampler

On the land surface:

- Land disposal of solid and liquid waste
- Dumps
- Fertilizers and pesticides

Below the ground surface but above the water table:

- Leaking underground storage tanks
- Landfills
- Sumps and dry wells at brownfield sites

Below the water table:

- Well disposal of wastes
- Mines
- Exploratory wells and test holes

Factors that affect groundwater contamination include:

- Characteristics of soil and rock through which the liquid waste flows.
- Degradation processes (filtration, sorption, dispersion, microbial degradation, etc.).
- Hydraulics of fluid flow.
- Nature of contaminant, particularly its stability under varying conditions.

Cyclic fluctuations of groundwater quality occur due to a number of diverse factors. For this reason, a single round of groundwater data are inadequate for establishing site conditions. The fluctuations occur at all depths within an aquifer.

Water Quality Investigation

Investigations of groundwater quality often have two problems–costs are too high and the data obtained turn out to be inadequate. These investigations have played a significant role in litigation, and must be conducted with a high degree of care. Investigations can be regional, local, or for a specific site. The significance of results is likely to be thoroughly questioned at meetings and even in court.

Prediction is difficult because of two main factors:

- Data on contaminant parameters (composition, concentrations, rate of release, size of source) are difficult to obtain and often incomplete.
- Dispersion of a contaminant is difficult to predict (Figure 3-3).

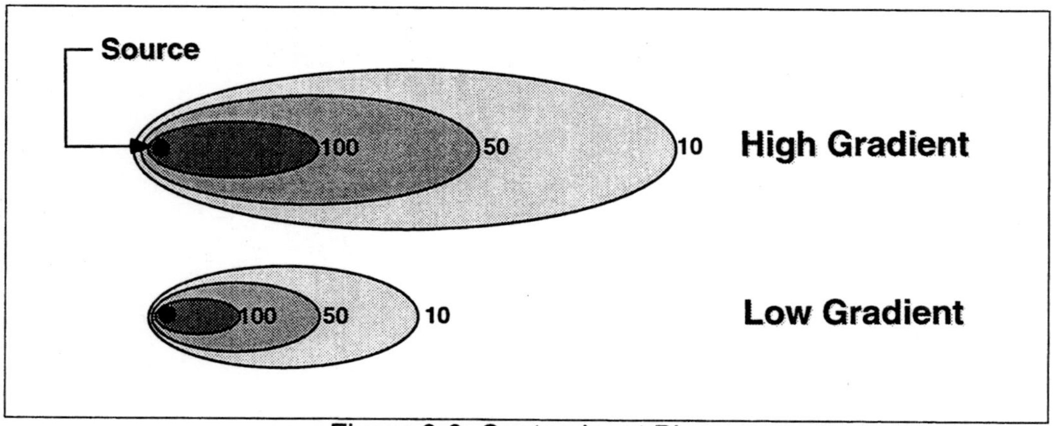

Figure 3-3. Contaminant Plume

Monitoring Well Design

The most important components of well design are the location and number of wells. The well placement should be such that the effect of a source can be monitored reliably and effectively. Most leaks or sources introduce substances into the ground that descend through the unsaturated zone, and upon reaching the water table begin moving in the direction of groundwater flow. Monitoring wells placed downgradient from a source will sample the contaminant plume effectively. Soil borings and preliminary monitoring wells may be needed to determine the depth and thickness of water bearing zone and the flow gradient. Therefore, often the groundwater investigations are phased efforts.

The number of wells is dependent on soil permeability, size of the source, and the size of the area being investigated. A minimum of three wells are needed to determine the gradient. Typical well installation is shown in Figure 3-4.

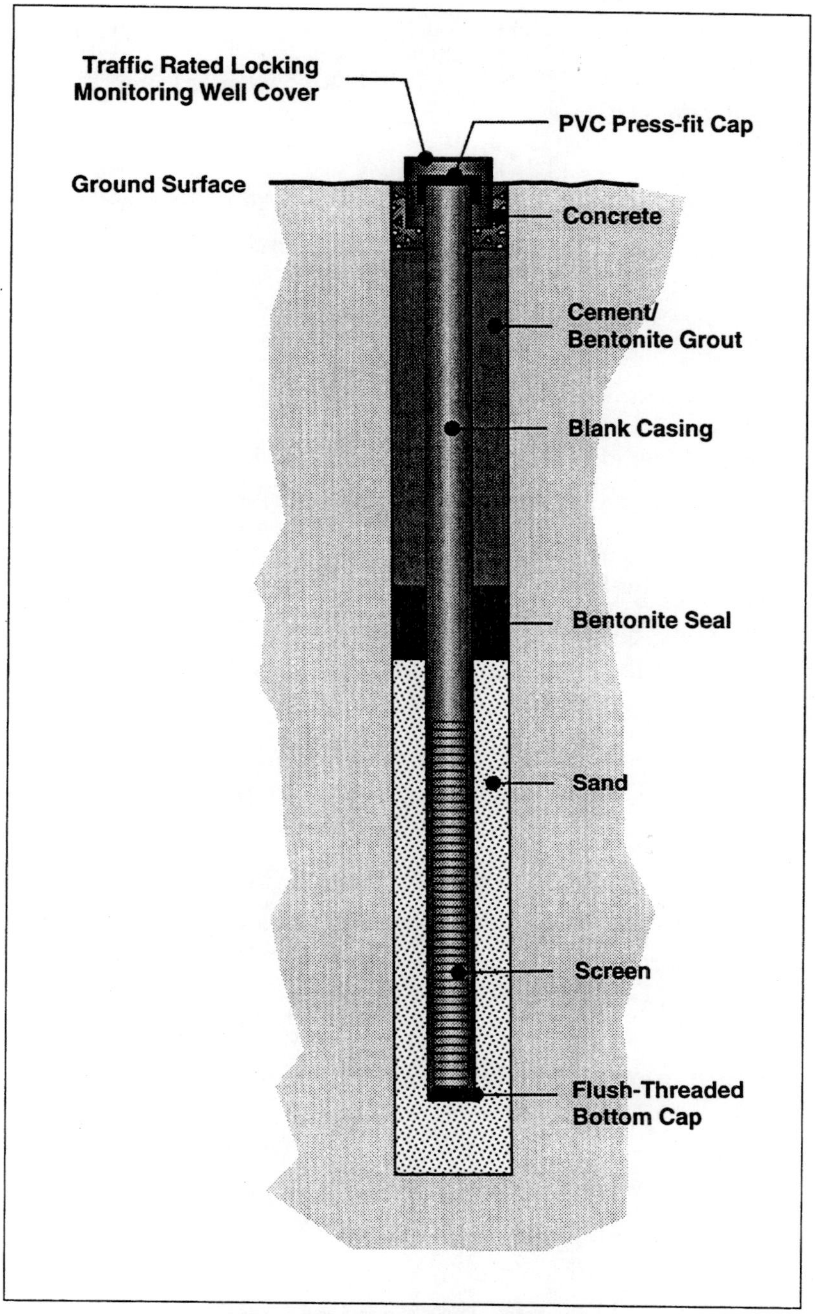

Figure 3-4. Monitoring Well

Diameter

Unlike water supply wells, monitoring wells are often installed in formations with low permeability. In deciding the diameter of a monitoring well, two considerations are important: 1)

Soil permeability—as a low-yield soil is difficult to sample when several well volumes have to be removed; 2) The sampling method—depending upon the depth, some sampling devices may not fit a small diameter well.

The amount of well water volume that may have to be removed, the recharge time, and disposal quantities for removed water are important considerations in deciding well diameter. The usual inside diameter range is from 2 inches to 4 inches.

Casing and Screen Material

Teflon is the best, but also the most expensive material. Stainless steel is the next best material, followed by PVC and Lo-Carbon or galvanized steel. Effects of contaminants on the casing material should be considered; for instance, PVC may be affected by chlorinated solvents, although it is very resistant to other chemicals. PVC primers and adhesives can release MEK and other organic compounds into the groundwater, leading to errors. If the contaminants of interest are already defined, and they do not include materials that might interact with casing, it may not be necessary to use the best and most expensive casing materials. The casing and well materials should be washed with a nonphosphate detergent and rinsed thoroughly with deionized water. Metal casings may need to be steam cleaned to remove cutting oils and lubricants.

Screen slot size is selected on the basis of grain size of the material surrounding the screen. If the natural soil in the formation is a uniform granular material, a screen can be placed directly against it. More commonly, a gravel pack is needed. The size of the screen (width of slots) and gravel gradation are selected based on filtering principles.

Screen Length and Depth of Placement

The full thickness of a water bearing zone should only be screened if the aquifer is being used for water supply. Otherwise, the screen length should be as short as practical to achieve the monitoring objectives. In detailed plume geometry studies, nested screen lengths of 1 to 2 feet are used (Figure 3-5), but for most site investigation, this approach would be unnecessary and also prohibitively expensive. However, the screen length should be limited to a length of 5 to 10 feet and should not extend above the water table to prevent:

- cross-contamination between layers.
- aeration when monitoring volatile organics.

Sealing Materials and Procedures

A monitoring well should sample a specific depth interval and be sealed off from the other layers. Therefore, effective seals are necessary just above the screened intervals and near the ground surface. Bentonite and cement grout have been used; bentonite does not shrink upon drying, but may be pervious to some organic liquids. Cement grout may shrink but is impervious, and when mixed with bentonite may provide the most stable barrier. Both cement and bentonite can introduce contaminants into the groundwater–cement by leaching chlorides and other by-products of hydration, and bentonite by ion exchange with chemicals in the well water. Therefore, when placed, they should not be too watery, so as to prevent migration into the gravel pack.

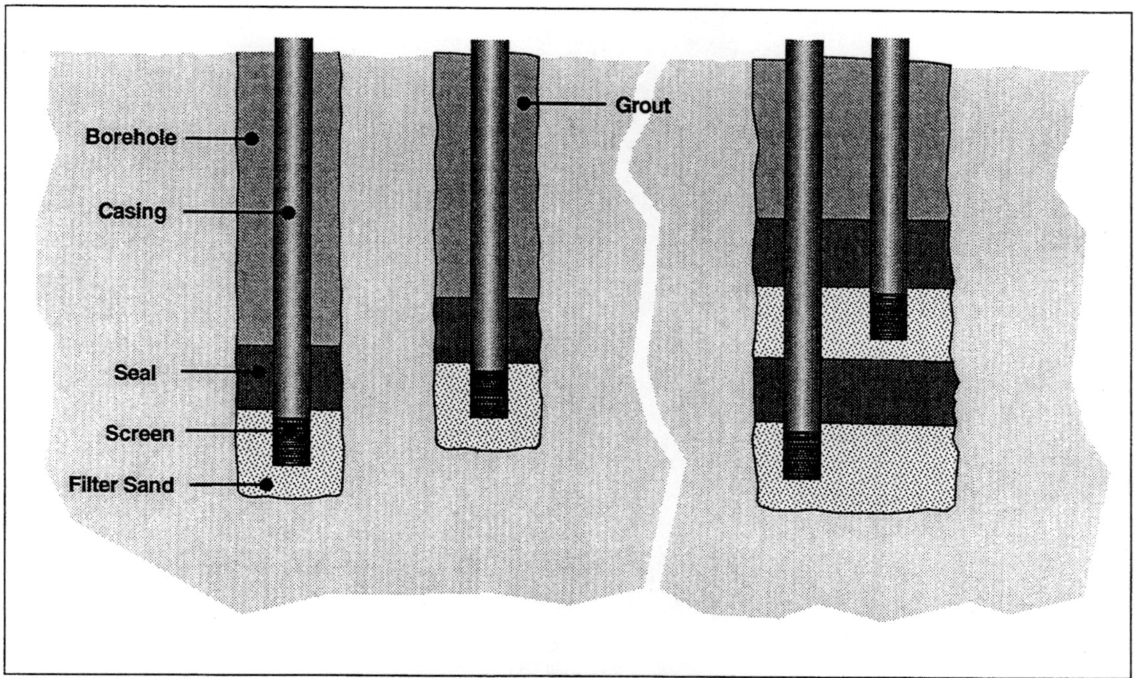

Figure 3-5. Multiwell Installations

Development

Drilling produces a smear of fine grained material along the borehole walls, and this "mud-cake" must be removed by flushing during well development. Additionally, development is needed to produce water free of suspended solids for sampling. Methods include surging or flushing with air or water as shown on Figure 3-6. In each method, it is intended to produce reversals of flow in and out of the well. A bailer can also be used by dropping and lifting rapidly to cause a back and forth flow of water in the well. Surge blocks or plungers, when moved vigorously up and down can move larger quantities than bailers, but can potentially damage a screen. Brief overpumping can be used in productive aquifers, where a high capacity pump is alternately turned on and off.

Pumping compressed air using a nozzle pointing into the screen is very effective, but the air must be filtered to capture oils or lubricants from the compressor, and workers at the surface must be protected, if large quantities of volatiles are present in the well.

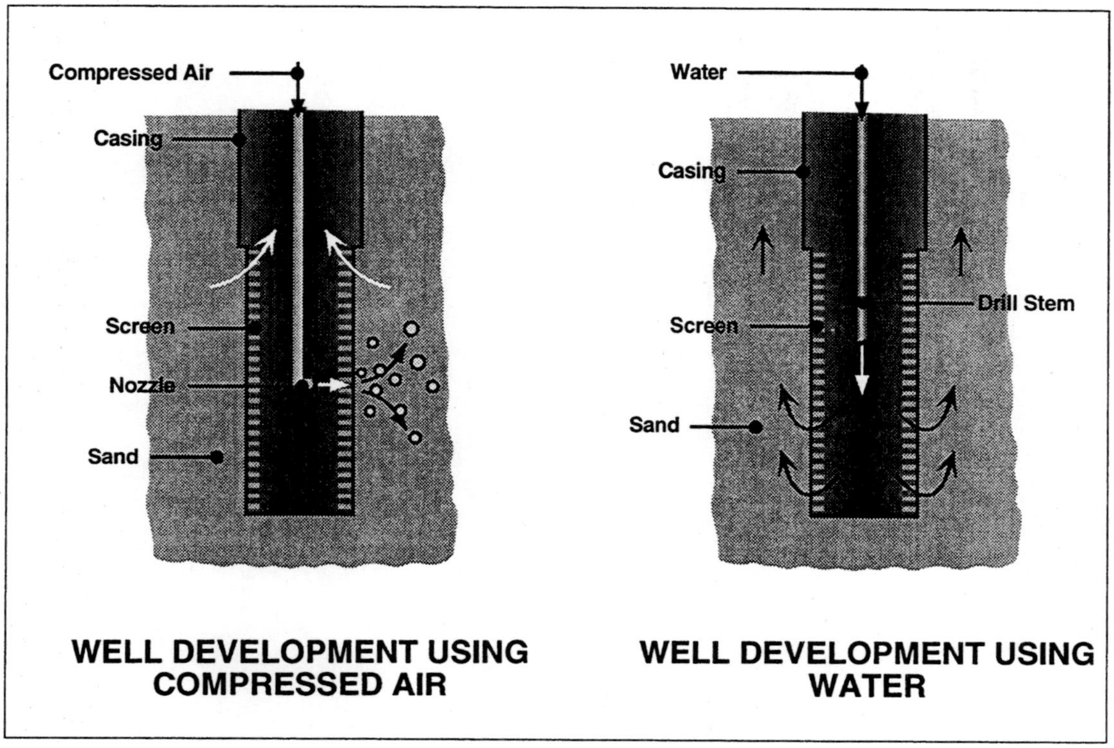

Figure 3-6. Well Development- Two Types

Development is difficult in low permeability materials, and if caving can be prevented, consideration should be given to development in an open borehole before the installation of gravel pack to prevent clogging.

Well Water Sampling

Wells used for sampling should have lockable caps for security. The first step after opening the lock is to check the water level. Then stagnant water, at least three to five well volumes, is removed. The well pumping parameters are monitored until stabilized (these include pH, electrical conductivity and temperature), after which a sample is obtained, preserved, and sent for laboratory testing (Figure 3-7).

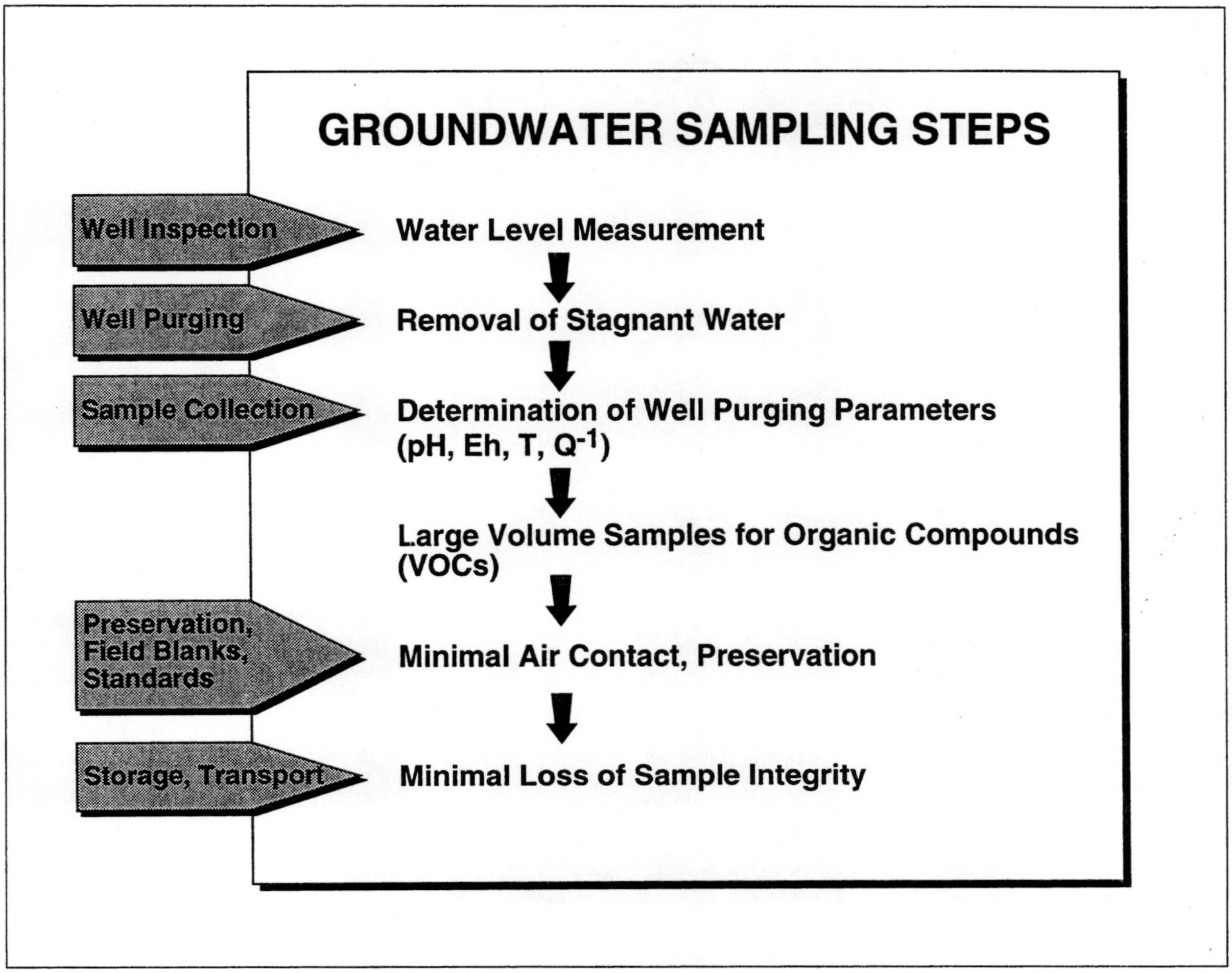

Figure 3-7. Groundwater Sampling Steps

3.8 ANALYTICAL TESTING

Portable Detection Instruments

A number of portable instruments are used for detecting off gases in the borehole or from soil samples. This is done for three purposes: 1) screening for health and safety monitoring for selection of appropriate level of protective equipment; 2) to get a preliminary scan of the presence of volatiles, which along with odor and visible staining, forms the basis of selecting samples for testing; and 3) for filling in data when drawing plume boundaries of contaminated areas.

Instrument	Manufacturer	Principle	Comments
OVA™	Century Systems	Hydrogen Flame ionization	Calibrated to Specific Organics
PID™	HNU	Photo Ionization	Specific lamp required for groups of compounds
Micro TIP™	Photovac	Photo Ionization	UV Lamp
FID	Sensidyne	Flame Ionization	Same as OVA
Multi-gas Personal Monitor	Bacharach	Various	Detects Oxygen deficiency, Hydrogen Sulfide and Carbon Monoxide

Quality Control

The analytical chemistry test methods in use for hazardous wastes have been selected and adapted to comply with Superfund guidelines. These methods include procedures for field and laboratory quality control, sampling, determining hazardous characteristics of wastes (toxicity, ignitability, reactivity and corrosivity), and analytical testing of chemicals adapted for hazardous wastes.

Laboratories certified by individual states are required to maintain QA/QC programs. In addition, on important investigations, project-specific QA/QC plans are prepared, addressing the particular sampling and analysis programs. Essential elements of a Project QA Plan include: Project Description and Organization; QA Objectives; Sampling Procedures; Analytical Procedures; Internal QC checks and Corrective Action Procedures.

Definitions

The following terms are useful in describing samples obtained for analytical testing:

Accuracy: Nearness of a measurement to the true value.

Precision: Agreement between a set of replicate measurements without knowledge of the true value.

Analyte: A chemical or waste compound that is the subject of analytical testing.

Surrogate: An organic compound that is similar to the analytes but not normally found in environmental samples, spiked into all blanks for calculating percent recovery.

Blank: An organic or aqueous solution that is as free of analyte as possible. Three types are commonly used:

Equipment Blank: Opened in the field, poured over or through the sample collection device as a check on equipment.

Field Blank: Transferred from one vessel to the other in the field as a check or environment.

Trip Blank: Transported to the field and returned without opening as a check on transport procedures.

Holding Times

The quality of even a well preserved sample begins to deteriorate with time, until it is no longer representative of the site. The following holding times are considered the longest intervals for which the sample quality remains acceptable. As a rule, the samples should be tested soon after retrieval.

Test	Container	Preservation	Maximum Holding Time
Bacterial Tests	Polyethylene (P), Glass (G)	Cool, 4°C	6 hours
Inorganic Tests	P, G	None for Chloride or pH; Others; Cool, 4°C	7 days to 28 days
Metals	P, G	Cool, 4°C	Chromium VI = 24 hrs. Mercury = 28 days Others = 6 mos.
Organic Tests	G, Teflon-lined Cap	Cool, 4°C	7 days for extractions, 14 days total time

Extraction Methods

A number of extraction procedures are available to separate the analytes from the matrix. Acid extraction procedures involve refluxing (flowing back) of a mixture of acids (nitric acid and hydrochloric acid) and the material to be analyzed. A good laboratory can select a suitable extraction procedure depending upon the analyte and the character of the sample.

For the analysis of volatile organics, a purge-and-trap method is needed. In this method, an inert gas is bubbled through the solution at the ambient temperature. The resulting vapor is swept through a sorbent column, which is later heated and backflushed with inert gas, thus desorbing the components onto a gas chromatographic column.

Laboratory Methods

Method Group 6010: *ICP (Inductively Coupled Plasma-Atomic Emission Spectroscopy)*

This group of methods is applicable to trace elements including metals in a solution. All sample types (sludges, soils, groundwater) require digestion before analysis. The method measures element–emitted light by optical spectroscopy. The spectra are dispersed and the intensities of the lines monitored.

Method Group 7000: *Atomic Absorption Methods*

In these methods, also used for metals, a sample is atomized in a flame; a light beam is directed through the flame onto a detector that measures the amount of absorbed light. The wavelength of the light beam is characteristic of the metal being analyzed, and the light energy absorbed by the flame is a measure of the concentration of that metal in the sample. Each method in this group analyzes a different metal, e.g.: 7020-Aluminum, 7061-Arsenic, 7190-Chromium, 7420-Lead, 7950-Zinc, etc.

Method Group 8000: *Gas Chromatography*

Also known as Vapor Phase Chromatography, it is useful for organic compounds capable of being volatilized without being decomposed or chemically rearranged. The method employs a column of absorbent material in a tube. Using helium as an eluant, the analyte sample is passed through the column. The individual substances in the sample differ in their tendency to be absorbed and so they move through the column at varying rates.

8010: Halogenated Volatile Organic Compounds.

8015: Nonhalogenated Volatile Organics.

8015M: Modified for fuel-range hydrocarbons (Gasoline, Kerosene, Diesel).

8020: Aromatic Volatile Organics (included Benzene, Toluene, Ethylbenzene, and Xylene.

8080: Organic Pesticides and PCBs.

Method Group 8200: Gas Chromatography/Mass Spectroscopy

Mass spectroscopy is used to detect the compounds separated by gas chromatography. The substance to be analyzed is ionized, the ions separated in a magnetic field, scanned, detected, and recorded.

8240: Volatile Organics (including Aromatics)

8270: Semivolatile Organics

PART II

SITE REMEDIATION

CHAPTER 4

REMEDIAL ALTERNATIVE SELECTION

4.1 SELECTION PROCESS

Choice of remediation method is the most important and expensive decision that is made regarding a contaminated site. During investigation any decisions made regarding methodology, choice of tests and the frequency of sampling can be updated and extended after a review of results. However, work done during remediation, if inadequate, may have to be scrapped and repeated, rather than extended and improved. Choice of remedial method commits a site owner and the project manager to installation expense and a course of action that may have to be followed for a considerable period and result in continued operational expenses. Selection of a remedial alternative can be made in an informal manner, or in accordance with formalized Feasibility Study methodology. Actual implementation of a remediation scheme at a site is sometimes held up due to the coordination required. For instance, a regional groundwater treatment scheme necessitates that pump-and-treat be initiated at contiguous properties simultaneously. Community factors such as the timing of academic year may need to be considered before beginning excavation of contaminated materials upwind from a school. Other legal, political, and financial factors may also influence the implementation sequence for a remediation scheme. Even if the technical project manager undertaking the remediation does not have any control over the factors that may delay implementation, these factors need to be anticipated and recognized by the project team to ensure smooth implementation of the engineering objectives.

The following paragraphs provide guidance on the technical, regulatory and financial interplay that takes place in the selection of remediation methods. Individual factors that are the basis of selection are discussed in separate paragraphs.

4.2 CONTAMINANT CHARACTERISTICS

The nature and distribution of contaminants are the foremost criteria for selection of remediation method at a site. In addition to the chemical properties, physical characteristics such as viscosity, solubility, and density become very important for collecting and treating a contaminant. The chemical properties and toxicity may be the main reasons why remediation is needed, but the physical properties may control the selection of a remedial method.

There are three main contaminant groups that typify the characteristics important for remediation: Volatile Organic Compounds, Semivolatile Organic Compounds, and Inorganic materials including metals. Additional related groupings such as chlorinated VOCs and halogenated semivolatiles have similar physical characteristics but higher toxicity.

Remediation techniques include three main steps: access, collection, and treatment. Access to a contaminant that is distributed in the subsurface is primarily accomplished in one of two ways: drilling wells or tunnels into the subsurface, or, excavating the contaminated materials to bring it to the ground surface for treatment. After the installation of vertical wells, or horizontal wells in trenches or above-ground stockpiles, collection is accomplished by various extraction processes such as application of a vacuum or pumping. Treatment of the contaminant involves concentration, adsorption, recycling, and incorporation into an inert mass or conversion into simpler and less harmful form.

For each of the three steps in the remediation process, the physical character of the contaminant governs the choice of remediation method, as described below.

VOCs

- *Access*: The volatile organic compounds generally spread over broad areas. Often, wells or access tubes are used to reach the contaminated zone.

- *Collection*: The VOCs have to be removed from the subsurface using some transporting agent such as air or groundwater. If groundwater is the transporting medium, then stripping is required for separating the volatiles after collection.

- *Treatment*: For VOCs, the treatment method is dependent upon their concentration at the end of collection. Fuel hydrocarbon at high concentrations can be recycled or burned with appropriate emission controls. Volatiles low in concentration and octane value, such as solvents, are condensed by using an adsorping medium such as Granular Activated Carbon, which is then sent for regeneration or disposal.

Case History: Jet Fuel at Coastal Dune Site

This case history pertains to a coastal airport site founded on sand dunes. The stratigraphy at the site consists of coarse sands to a depth of over 100 feet below the ground surface. The water table is located at about 40 feet depth below the ground surface. A tank farm near the airport terminal consists of several large above ground tanks that store jet fuel. Refueling areas on the aprons near the gates are supplied by underground piping from the tank farm.

Over the years the above ground storage tanks leaked, and a significant amount of jet fuel accumulated in the ground. Most of the jet fuel trickled straight down in the coarse sand layers and reached ground water at the depth of about 40 feet. Investigations showed that up to 1foot of free product was floating on the water table. Studies were underway to investigate the lateral extent of the problem and design a remediation scheme.

It was at this stage that redevelopment and expansion began at one of the airline terminal buildings. This work included excavation of foundation trenches in the apron area for new gates, jetways and pedestrian tunnels. The footing excavations unearthed several abandoned fuel supply lines, which had clearly leaked in the past, leaving behind soil staining and hydrocarbon odors. A health and safety plan was drawn up for the construction workers who were doing the excavation and construction.

As construction progressed, the trench excavations extended to areas beyond where fuel supply lines had existed in the past. The soil was not stained and there was no bent and rusted metal tubing or other remnants of fuel supply pipes. However, strong hydrocarbon odors continued, and high concentrations of organic vapors persisted in the trenches. This was a source of concern as there were health and safety restrictions on workers, and they could not enter the trenches without respiratory protection when high vapor measurements approached 100 ppm. A brief investigation was performed to search for the source of vapors. Jets taking off from the nearby runways created a lot considerable amount of jet exhaust which had a high hydrocarbon content. However, their influence was only on the ground surface, and subsided soon after take off.

Sampling was performed in trench sidewalls to compare the organic vapor measurements with fuel concentrations in the soil. Soil samples obtained near the high OVA measurements were found to contain no detectable concentrations of hydrocarbons (Figure 4-1) It became clear that the vapors were travelling up from the free product floating on the water table, through the porous coarse sand and into the trench. Presence of pavement or a dry crust on the sand dunes was preventing widespread transmission of vapors, and freshly excavated trenches provided a convenient escape route. The case history illustrates ease with which organic vapors travel through porous media, and that caution should be shown in equating vapor concentrations with concentrations in the nearby soils.

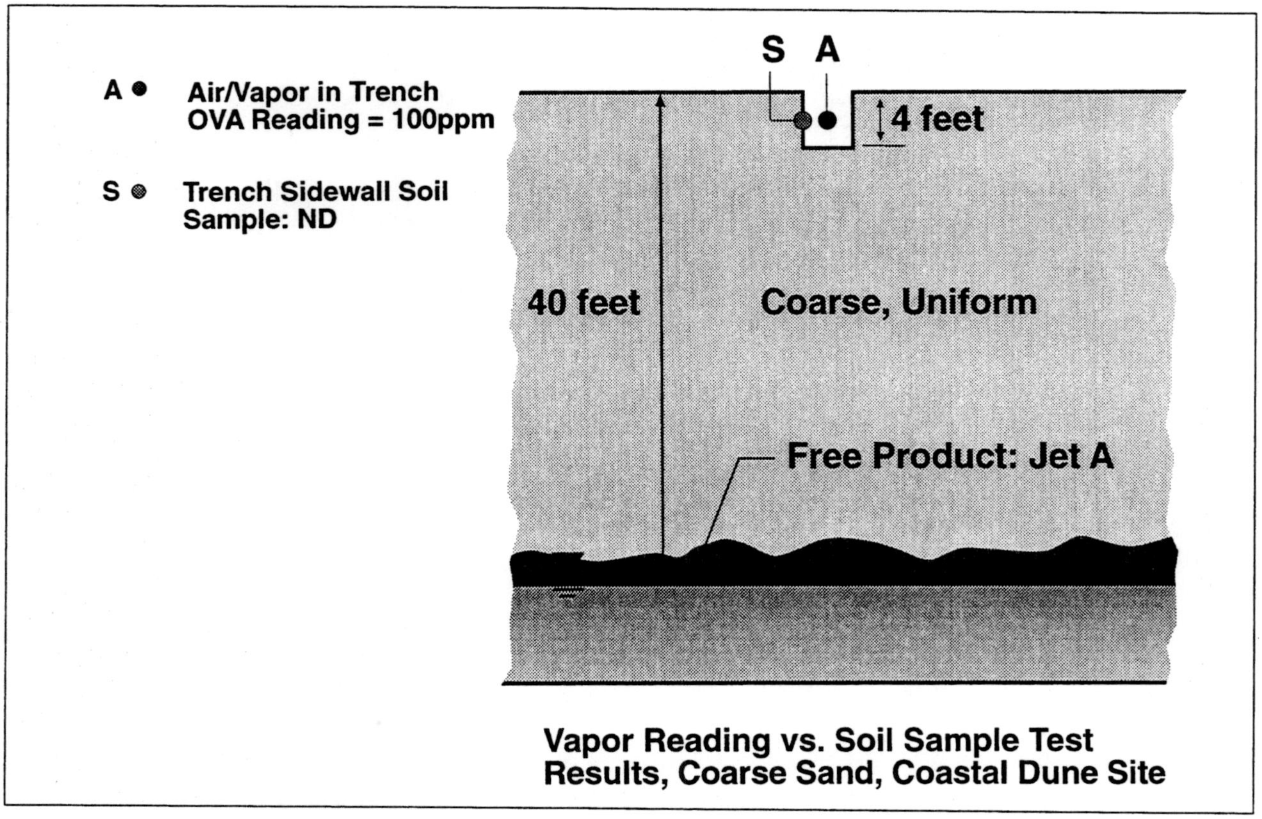

Figure 4-1. Vapor Reading vs. Soil Sample Test Results, Coarse Sand, Coastal Dune Site

Semivolatile Organic Compounds

- *Access*: Excavation and stockpiling of soils contaminated with semivolatiles is the most effective method of gaining access to the contaminants. Physical site restrictions such as the presence of buildings may preclude excavation and thus require the use of wells.

- *Collection*: An aggressive flushing agent such as steam or a solvent may be required for separating the contaminant from the soil matrix for collection purposes. The contaminants may be so firmly bonded to the matrix as to require agitation in the presence of a detergent or solvent.

- *Treatment*: Options for treatment are generally limited to landfill disposal, recycling, or asphalt incorporation.

Case History: Towne Gas (Manufactured Gas) site in Venice, California

Venice, California is a seaside community where a manufactured gas plant operated from 1905 to 1914. Cracking crude oil or coal at high temperatures in the presence of steam produced the manufactured gas, or towne gas. The gas, which consisted to lighter fractions and resembled petroleum gas or propane, was used for street lighting. As natural gas became available in the early part of this century, these plants began shutting down. The plant in Venice was operated for only nine years before it became obsolete. As a byproduct of the gas-manufacturing product, these plants produced a number of wastes and usable petroleum products, all of which were stored or disposed of on the site, either in shallow brick-lined pits. The wastes and byproducts were thick viscous or powdery substances that included lamp black and several carcinogens like Benzo (a) Pyrene.

Even after the manufacture of gas ended, the utility company maintained the site as a distribution center for natural gas, and utilized the large above ground steel tanks for storage. The plant yard was converted to a vehicle servicing facility, and a couple of fuels supply pumps and underground gasoline and diesel storage tanks were installed. Over the years these tanks leaked, and permeated into the waste pits and transported the viscous byproducts of gas manufacture down trough the coarse beach sand down to the water table at 25 feet depth. Gradually the site just became a utility vehicle facility, and surface remnants of gas manufacture or storage were removed. The property was subdivided and sold off by the utility in the 1970's. One parcel remained undeveloped for another ten years, and was then selected by an advertising agency as the site of their new headquarters.

As excavation for site clearance began, it was clear that this was no ordinary site, as construction equipment encountered accumulations of thick gooey, odorous materials. Residential and commercial neighbors at adjacent properties complained, and the grading was halted and a site investigation was performed. The site investigation defined a viscous plume of various viscous waste products floating, or rather sitting on the water table (Figure 4-2). A slight gradient was detected in the ground water, which was close enough to the beach, about ¾ miles, that it showed a twice-daily tidal influence of about 6 inches variations in elevation. However, even after eighty years the wastes had not migrated off site, and ground water in wells surrounding the site did not contain detectable amounts of the contaminants present at the site.

As it became clear that the wastes at the site had come from the utility operations, the appropriate utility companies stepped in and assumed responsibility for the remediation at the site. At this point, several divergent viewpoints emerged. The utility companies wanted site closure with minimal future liabilities. The owners wanted their dream buildings constructed. The adjacent property owners were concerned that their parcels may also be contaminated, and wanted to be compensated but not disturbed. the surrounding community wanted to remain free of health risk, and the elected officials wanted to lead the community. The construction contractor wanted to build the building. *(continued on next page)*

A number of open public meetings were held to obtain input. In addition, a number of technical and legal meetings took place. What finally emerged after about a year of debate was a compromise that satisfied almost everyone. The building concept included a three-level underground parking garage, necessitating an excavation. Therefore, any in-situ remediation schemes such as isolation, containment or capping were not feasible. Due to the viscous nature of contaminants, any in-place treatment or extraction would have required steam injection or solvents. In-place stabilization using methods such as deep soil mixing with cement or other pozzolanic materials also would not have provided the subterranean space.

The remedial alternative selection choices were limited to the various methods of excavation and off-site disposal. Much of the engineering effort was used for devising a safe, economical, and practical scheme that was acceptable to the community. The excavation plan was presented to the community in public hearings, and modified a number of times. The final remediation scheme is described in a case history in Chapter 6 and the site operations illustrated in Figure 6-5.

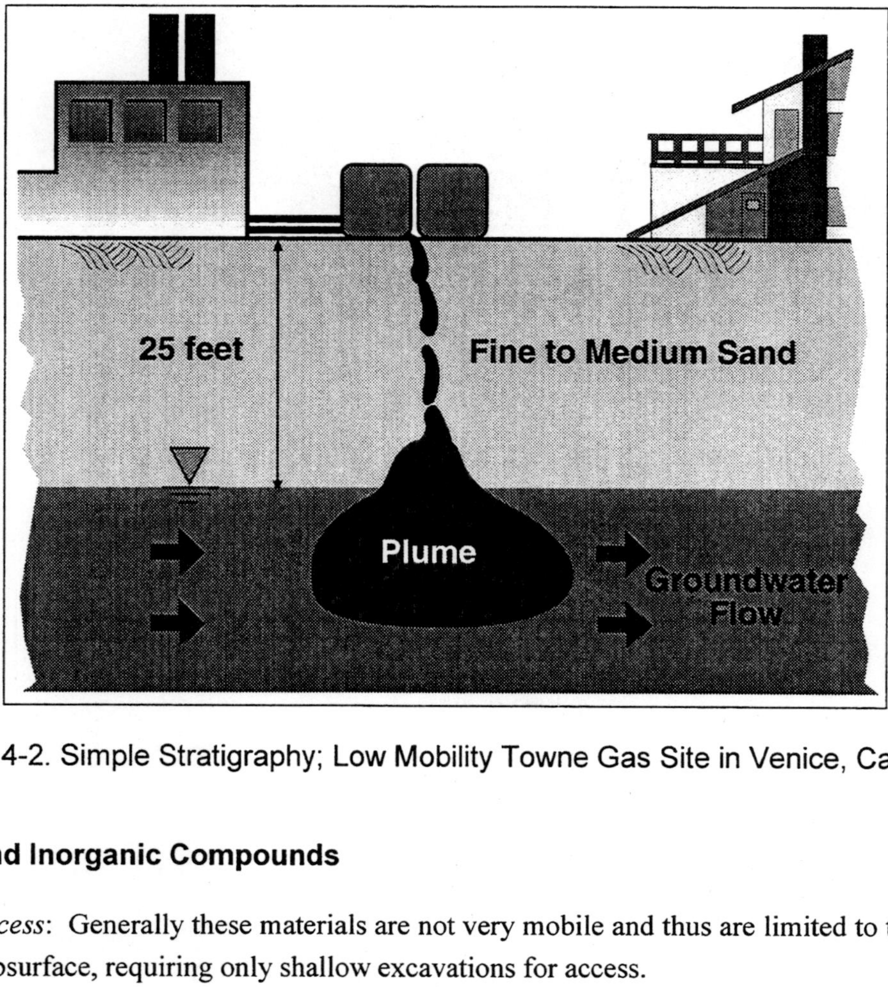

Figure 4-2. Simple Stratigraphy; Low Mobility Towne Gas Site in Venice, California

Metals and Inorganic Compounds

- *Access*: Generally these materials are not very mobile and thus are limited to the shallow subsurface, requiring only shallow excavations for access.

- *Collection*: It is extremely difficult to separate the metal or pesticide contamination from the soil matrix. Soil Washing with aggressive agents is the only method that is widely attempted.

- *Treatment*: Stabilization is the preferred option, followed by landfill disposal. Metals that have been concentrated by soil washing, or the soil and contaminant right after excavation can be stabilized.

Case History: Farmland Site in Central Valley of California

At a farmland site at a riverbank in the Central Valley of California, a small shack was operated as a car battery dismantling operation. After removing usable parts for sale, the carcasses of hundreds of batteries were left behind. Suburbs crept ever closer to the farmland, and in the mid-1980's the site was considered for residential development. Site investigation discovered total lead concentrations all over the 200-acre site, with some samples reaching concentrations of over 200 ppm. When testing was performed for soluble lead, it was found that only the battery shack and a three-acre area showed high concentrations, as high as 50 ppm. Soluble lead was not detectable in samples obtained from the remaining 197 acres (Figure 4-3).

Based on background levels in the region, the total lead was determined to be a natural occurrence, and only the soluble lead in the three acres area was required to be remediated. The site was excavated down to about 10 feet depth, and the soil was stabilized using a pozzolanic admixture. Performing bench-scale treatability studies, followed by leaching tests first proved the method. A pugmill was used to blend the soils and the pozzolanic additive. The resulting stabilized material had a consistency that was suitable for use as backfilling the excavated area. Monitoring wells were installed to check if any lead was leaching into the groundwater, and could threaten the river water quality. Monitoring for two years showed no sign of lead in the groundwater.

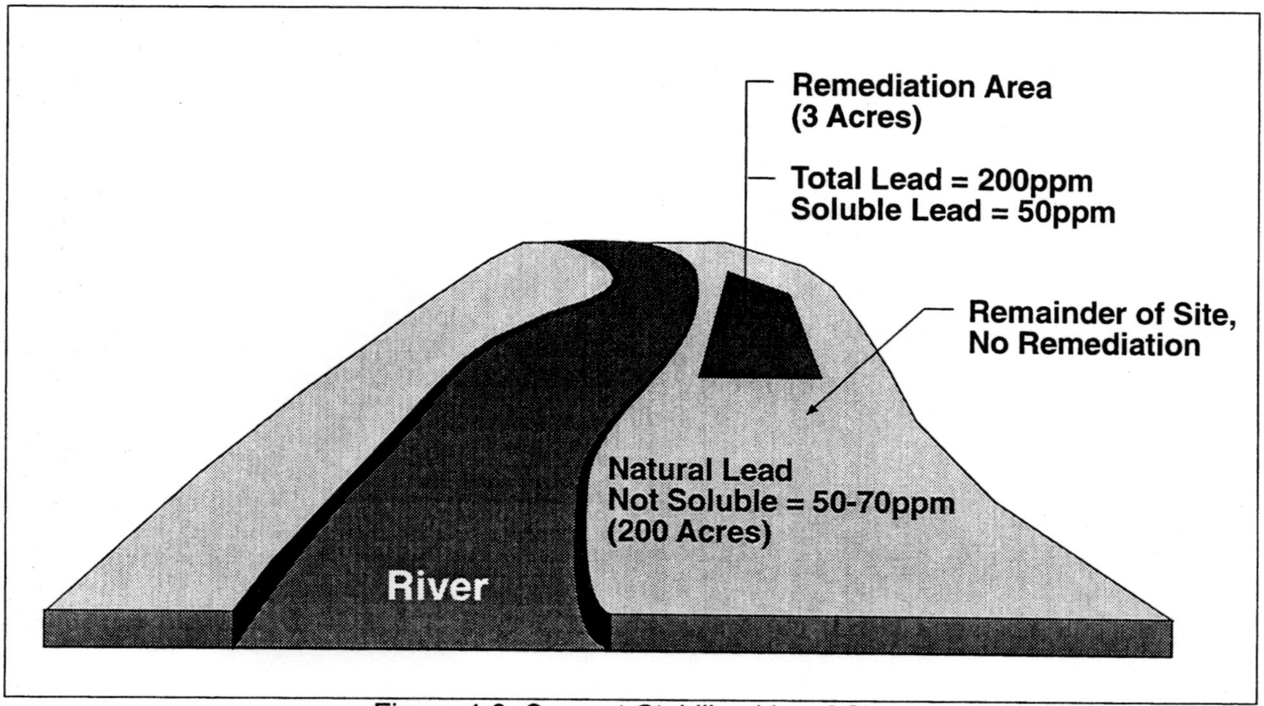

Figure 4-3. Cement-Stabilized Lead Site

4.3 PHYSICAL SITE SETTING

Choice of remediation method is influenced by the physical setting of a site in several different ways.

Presence of Buildings

If the ground surface at a site is occupied by buildings and contaminants are present in the subsurface remediation methods will need to be adapted to the site. Slant wells, horizontal trenches, directional drilling, and microtunnelling can be used to reach zones inaccessible from the surface.

Surrounding Community

Presence of sensitive populations (school children, senior citizens, hospital patients) in the surrounding community may preclude the use of on-site handling and treatment. Even if space is available at a site, any ex-situ methods such as land farming may be impractical. Emissions of volatiles, dust, and noise may not be permitted. Health risk and community disruption due to traffic and heavy equipment operations may preclude certain remediation choices.

Soil Type and Groundwater

The local stratigraphy at a site, and the presence of groundwater is an important factor in selecting remediation methods. Vapor Extraction Method is ideally suited for coarse sand in the vadose zone and would be ineffective at a massive clay site with a shallow water table. If excavation is to be performed, the in-place strength will determine the slope angle at which the excavation sides can be cut. The need for pretreatment processing and the suitability of material-handling equipment such as screens, shakers, conveyor belts, and slurry tanks is a function of soil grain size at the site.

4.4 COST AND SCHEDULE CONSIDERATIONS

Even if cost and schedule are not explicitly considered in selecting a remediation method, the implementation may only proceed in accordance with an owner's ability to pay for site work. The selected remediation method may need to be broken into stages, or may need to be implemented in a phased manner, beginning with a pilot test, and then gradually on to full-scale implementation. Industrial owners who intend to maintain ownership and land use, may choose to leave contaminated materials on site and just perform sufficient remediation for containment or isolation to limit the spread of contamination beyond site boundaries.

Opposite priorities may be in effect at sites which need to undergo change in ownership or land use, necessitating a rapid "dig-and-haul" approach so a complete closure could be obtained and the proposed sale put through. Even if other, slower techniques are more cost effective or technically superior, schedule will control the choice of remediation method if property transfer is desired.

4.5 REGULATORY REQUIREMENTS

If remediation is performed under court-mandated or regulatory agency-required guidelines, the details of regulations take on paramount importance. Even the specific method of remediation may be prescribed in the court order. Although the remediation method may have been initially chosen on the basis of good engineering, once written into a legal document, it may be very difficult to change even if conditions change or newer methods become available.

Certain individual chemicals, if they have leaked from a fuel storage tank can be dealt with in a straightforward manner in accordance with LUST regulations. On the other hand, similar chemicals, if leaked out of a waste storage tank, may require a much more elaborate RCRA closure. Regulations that allow the use of health-risk based cleanup criteria, may allow different residual concentrations at different locations with different surroundings. Different states or local jurisdictions have different criteria for cleanup of similar chemicals, and their laws may vary for similar contaminants such as gasoline from underground fuel storage tanks.

UST investigations for a single tank at a gas station provide a simple model for scoping of site investigations. If the tank is to be removed, often little more than two confirmatory soil samples from below the tank invert level require testing. If contamination is detected, the investigation may need to be expanded to include drilling and sampling. Remediation methods and their extent may depend on the detail with which investigation is performed.

Required Investigation: Local Regulation Comparison				
	Los Angeles, CA	Chicago, IL	Cranford, NJ	Miami, FL
Case A-Tank Removal: samples required from tank excavation	▪ 1 beneath each end of tank invert.	▪ 1 beneath each end of tank invert. ▪ 1 from each 20 ft. of excavation sidewall at a depth representing the lower 1/3 of tank. ▪ 1 every 20 ft. beneath piping.	▪ 1 every 5 ft. along tank centerline. ▪ At least 2 from excavation side- walls.	▪ 1 groundwater ▪ 4 OVA
Case B–Tank Has Leak: detection investigation	▪ 2 borings to 40 ft depth.	▪ 4 borings to 15 ft. depth (common practice, not a requirement).	▪ 8 borings to 40 ft. depth.	▪ 1 boring to 5 ft. below ground-water table (approx. 10 ft. below ground surface)

4.6 TECHNICAL FEASIBILITY EVALUATION

The formal Feasibility Study methodology developed for Superfund sites provides a framework in which to weigh alternate schemes available for remediation. At some sites, the structured RI/FS (Remedial Investigation/Feasibility Study) methodology can be used to select a suitable remediation scheme. Large, complex sites under public scrutiny are prime candidates, as the process allows a methodical comparison of very dissimilar factors.

RI/FS studies for Superfund sites have become known best for producing volumes of paperwork rather than actual cleanups. Even if such a formal feasibility evaluation is not performed at a site, a conscious effort should be made to develop and screen alternates. The following steps are suggested:

- Develop objectives for remedial action, specifying whether soil and groundwater contamination or air emissions are to be treated.
- For each of the media—soil, groundwater, air—list the likely remedial steps such as excavation, pumping, capping, etc.
- Identify and screen the remedial technologies that may be applicable.

During screening, interactions between soil, groundwater, and air treatment processes are considered. Elimination of the source of contamination may be the only comprehensive answer. The overall process of development and screening is illustrated below with an example of a straightforward site where hydrocarbons and aromatic volatiles have impacted the site soils.

Environmental Media	Remedial Objectives	Remedial Step	Remedial Technology	Others Requirements
Soil	▪ Prevent ingestion/ direct contact ▪ Prevent inhalation ▪ Prevent migration	▪ No action ▪ Containment ▪ Excavation/ ▪ disposal ▪ Removal/ treatment	▪ Fencing/deed Restrictions ▪ Capping ▪ Removal/treatment	▪ Groundwater ▪ Air ▪ Structures ▪ Solid wastes

4.7 TREATABILITY STUDIES AND PILOT TESTING

Treatability studies are intended to assess the process as applied to the site materials—soils and the contaminants—and also allow optimization of the process by adjusting such parameters as nutrient proportions, temperature, and durations for the various steps involved in the treatment process. In contrast, the pilot tests are intended to assess the site, and optimize such site-specific parameters as the radius of influence, drawdown rates, etc. In certain situations both treatability studies and pilot testing may be required.

Any remedial method selected for final implementation should be tested on a limited scale before launching into full-scale treatment. Some of the processes such as biotreatment can be tested in the laboratory in bench-scale experiments termed treatability studies. Soil from the site is used in the laboratory tests with actual additives or treatment agents. Precise measurements of oxygen intake, carbon dioxide outtake, temperature effects, microbial growth, and contaminant consumption form the basis for planning a full-scale treatment. Other types of treatment methods can also be evaluated at bench-scale level, including soil washing, solvent extraction, and chemical treatment.

Pilot tests are small-scale tests of the remediation process at the actual site, and are required for in-situ schemes including vacuum extraction, pump-and-treat, and bioventing. Even ex-situ methods such as biopiles or biocells which are applied to contaminated materials after excavation can benefit from pilot tests for evaluating the feasibility and working out the proportions of nutrients, etc. The in-situ pilot tests may require a level of planning and site work that may be similar to the full-scale treatment. Thus, if funding or regulatory requirements are not a hindrance, the pilot test installation, if successful, may be incorporated into the full-scale treatment scheme.

Treatability studies and pilot tests by necessity are somewhat experimental, so processes can be fine tuned and any fatal flaws discovered at an early stage. Both are intended to save time and effort by fine tuning the actual full-scale treatment. However, neither should become an end in itself, nor a research project for developing a new process—unless such research is undertaken intentionally. However, the overall methodology has become known well enough that a number of commercial laboratories are available to perform the treatability studies. A number of remediation contractors are able to offer pilot scale testing services. Some of them may even have regional or basin-wide permits and portable treatment equipment that offers a short set-up time and convenient regulatory approval for site-specific testing.

4.8 METHODS OF REMEDIATION

A discussion of the methods of remediation requires that the methods be grouped in some manner. The remediation methods can be grouped in two main ways: (1) According to the manner in which the contaminated zone is accessed, namely, in-situ remediation or ex-situ remediation, and, (2) By the nature of the treatment method, such as biological, thermal, chemical, etc. A third possible grouping scheme would be the nature of the impacted media. These include solids or near solid, mineral matter (soil, sludge, sediments, etc.), liquids (groundwater, industrial effluent, etc.) and air (soil gas, extracted vapor, emissions, etc.) or building materials (concrete, construction debris, lumber, vents, etc.)

The routine remediation techniques presented in the remainder of Part II are organized into two groups: Chapter 5 presents the in-situ remediation methods and Chapter 6 summarizes the ex-situ methods. Some of the innovative and integrated techniques are listed in Chapter 7.

CHAPTER 5

IN-SITU REMEDIATION METHODS

5.1 DESCRIPTION

In-situ remediation methods treat the contaminated materials in place, without large-scale disturbance or excavation of impacted soil. These methods are particularly suited to sites where surface disturbance would be disruptive to on-going facility operations because the impacted media lie beneath existing facilities or buildings. At sites contaminated with hazardous volatile contaminants, excavation could cause a release into the atmosphere; in-situ techniques can protect adjacent properties, such as residences or schools and their sensitive populations.

In most cases, some disturbance of the surface is required to install the treatment system. Penetration of the subsurface for contaminant collection can be accomplished by a host of drilling or excavation techniques, depending upon the contaminant type, plume depth, and the soil matrix. All in-situ techniques require considerable time for treatment and are unsuitable for use at properties slated for immediate sale or redevelopment.

In-situ methods may be slower, but only need a gradual expenditure of treatment expense. However, some site monitoring and agency reporting may continue for a longer period, and the final outcome of treatment may not be known for several months or even years. Some of the more popular treatment methods—such as vapor extraction—are particularly suited to the in-situ setting while others can be adapted for in-situ application by appropriately accessing the plumes. There are only a limited number of treatment techniques such as incineration that are impractical for in-situ application.

5.2 ISOLATION OR CONTAINMENT

In this remediation method, the area of concern is separated from the environment, but destruction of the contaminant is not accomplished except passively. Containment can be achieved by placing physical barriers in the path of contaminated media flow. Caps or covers are used to reduce surface water infiltration and cutoff walls or grout curtains can curtail groundwater inflow. Caps or covers are constructed from surface sealing materials, including synthetic membrane liners. These are made of polymers such as HDPE (High Density Polyethylene), have low permeability ($K<10^{-7}$ cm/sec), and must be compatible with the chemicals impacting the media. Liners are susceptible to damage by puncturing and tearing, and have to be sandwiched between protective layers of compacted soil or synthetic cushions.

Low permeability liners can also be constructed using clay, or by using a soil/bentonite admixture. A constructed cap of asphalt or concrete is also feasible for certain applications, and multilayer caps consisting of an upper protective layer, a drainage layer, and a clay layer are common.

Methods for reducing groundwater inflow include slurry walls that are constructed in trenches keyed into an impervious layer below the contaminated area. Grout curtains are formed by injecting grout through a series of pipes inserted into the strata to be waterproofed. Sheet piling driven into the ground can also form a physical barrier.

Isolation and containment methods have been used for many years. One site involved a slurry wall and cap in conjunction with a groundwater recovery system that achieved 90% efficiency. Design considerations include addressing site-specific conditions, diverting surface water, reducing infiltration, groundwater flow control, and consideration of future uses of the site. Equipment requirements are the same as for heavy subsurface construction and earthwork. Environmental effectiveness can be over 90%. Installation costs range from \$5 to \$30 per square foot of the cutoff area.

5.3 VAPOR EXTRACTION SYSTEMS

This is a process by which volatile compounds are removed from the soil by using forced air currents (Figure 5-1). This can be a very effective and efficient process for the right compounds and soil type, and many sites have been successfully remediated with its use.

The basic phenomena governing vapor extraction are simple. By applying a vacuum to a series of extraction wells, air flow is created through the vadose zone. Because of the pressure gradient created by the vacuum pumps, air moves through the unsaturated soil zone and then continues on into the perforated piping. The soil air containing vapors of volatile organic compounds flows through a liquid-vapor separator. The vapor is then treated using an emission control system such as activated carbon, catalytic oxidation, thermal oxidation, or internal combustion engine, etc. prior to its discharge to the atmosphere.

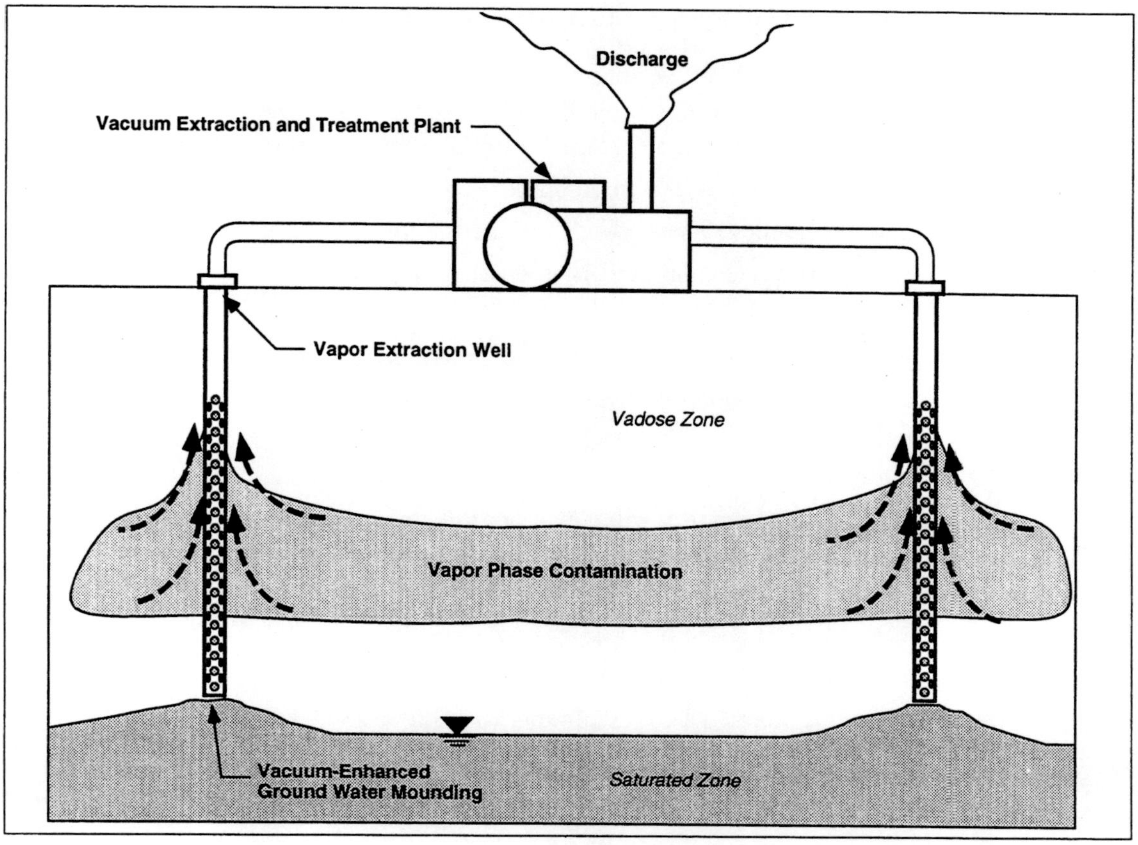

Figure 5-1. Soil Vapor Extraction System

The system can also include a series of air injection wells in addition to the air extraction wells. The purpose of the injection wells is to force atmospheric air down into the contaminated soil where it can collect VOCs before withdrawal through the extraction wells. The VOC-contaminated air can then be treated as stated above. Because of the common hardware, especially wells and ductwork piping, this method can be combined with some other techniques such as air sparging and bioremediation.

Although vapor extraction systems vary in size and design, the basic operation of an extraction and treatment system includes the following:

- A blower to act as a vacuum fan.
- A group of extraction wells with slotted screen pipes.
- A treatment unit to recover the volatilized compounds, often Granular Activated Carbon (GAC).

- Miscellaneous air flow meters, controls, valves, sensors, recorders, and control systems.
- Vent ductwork, piping, and heaters

The components of soil extraction system are basically off-the-shelf items; however, the design, operation, and monitoring of the extraction system require a certain amount of optimization.

This method is applicable to compounds with high Henry's Law Constants greater than 3 x 10^{-3} atm-m/mole—a measure of the equilibrium distribution between air and dilute solutions—used as an indicator of strippability. The method is effective for gasoline and other volatile and moderately volatile organics. There is usually a residual concentration below which the vapor extraction system is unable to economically remediate the site.

The effectiveness of a vapor extraction and treatment system is highly dependent on the site conditions. Factors influencing volatilization of hydrocarbons fall into four categories:

- Soil (high clay content, high water content, and low permeability reduce effectiveness)
- Environment (temperature, wind, evaporation, precipitation)
- Chemicals (Henry's Law)
- Well design and spacing

It can work successfully at sandy sites impacted with volatile hydrocarbons and no shallow groundwater, requiring about 12 to 18 months to achieve remediation goals. It is necessary to monitor groundwater (if impacted using wells), process flows, and exhaust gases both before and after treatment. It is difficult to monitor the progress of the remediation as the impacted media can only be tested by drilling and sampling.

Permitting agencies include the air management districts for emissions and the local fire departments for system integrity. Exposure pathways are virtually eliminated once the emissions are controlled. For a typical medium size system, the capital costs are about $100K to $200K including installation. Operations and maintenance is $1,000 to $5,000 per month with periodic carbon replacement at $10 to $20K. In qualitative ranking of costs, it falls in the middle.

5.4 AIR SPARGING

Groundwater pumping is commonly used for water supply (Figure 5-2), and water can generally be pumped directly from aquifer to the water supply. The natural filtration processes in the subsurface eliminate many of the contaminants and impurities that necessitate treatment of surface water supply from rivers or storage reservoirs. Groundwater contaminated with volatile organics can either be treated by conventional pump and treat methods or by air sparging.

Air sparging transfers the contaminants from water to the air by stripping of volatile organic compounds within the subsurface. Injection of air into the saturated zone can cause transfer of dissolved or adsorbed phase to the vapor phase. Once the VOCs reach the unsaturated zone, they can be extracted and then treated prior to release to the atmosphere. Air sparging, when combined with vapor extraction, precludes the need for expensive pump and treat systems (Figure5-3). The air injection area has to be carefully controlled to be within the vapor extraction influence zone. A good understanding of site geology and hydrogeology is needed to prevent spreading of contaminants.

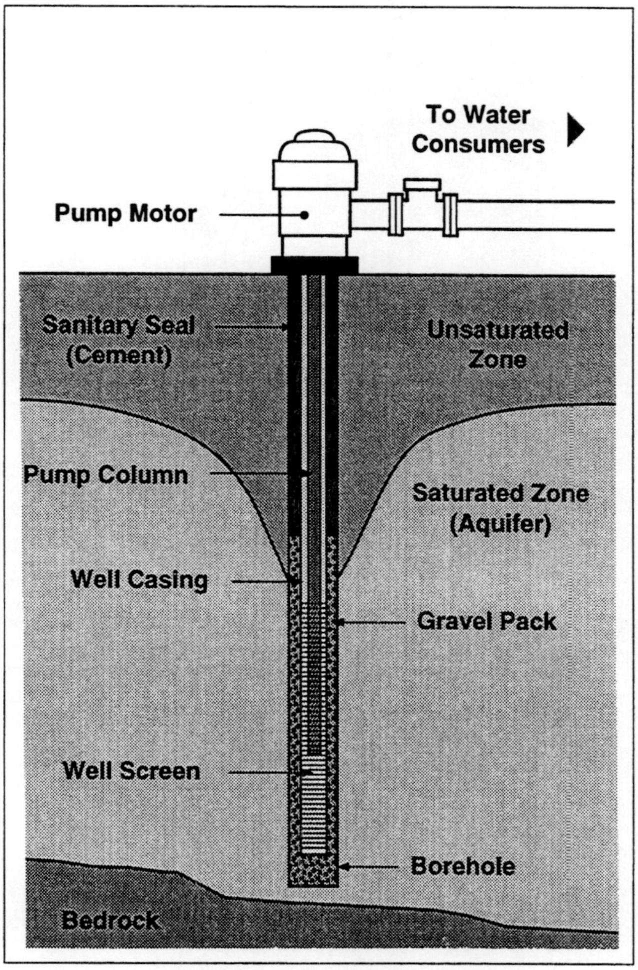

Figure 5-2. Groundwater Pump

The injection of air into the saturated zone also increases the oxygen concentration in the saturated zone—faster than through treated water reinjection or infiltration galleries. If there are no limiting parameters, the increase in oxygen concentration would likely result in enhanced bioremediation of organic chemicals. It has been reported that at dissolved oxygen concentration of 3.5 mg/l in the groundwater, the aerobic bioremediation appears to be most efficient.

One application of the air sparging technique is the use of sparge barriers. This technique involves a series of sparging points downgradient from a dissolved plume. Injection of oxygen through the sparging points creates an oxygen rich zone which encourages greater biological activity at the leading edge of the plume, where concentrations of chemicals are low enough to activate bioremediation. Consequently, the likelihood of chemical migration off-site is reduced.

Steam sparging involves injection of steam or heated air into the subsurface. This allows removal of some semivolatile organic compounds by heat.

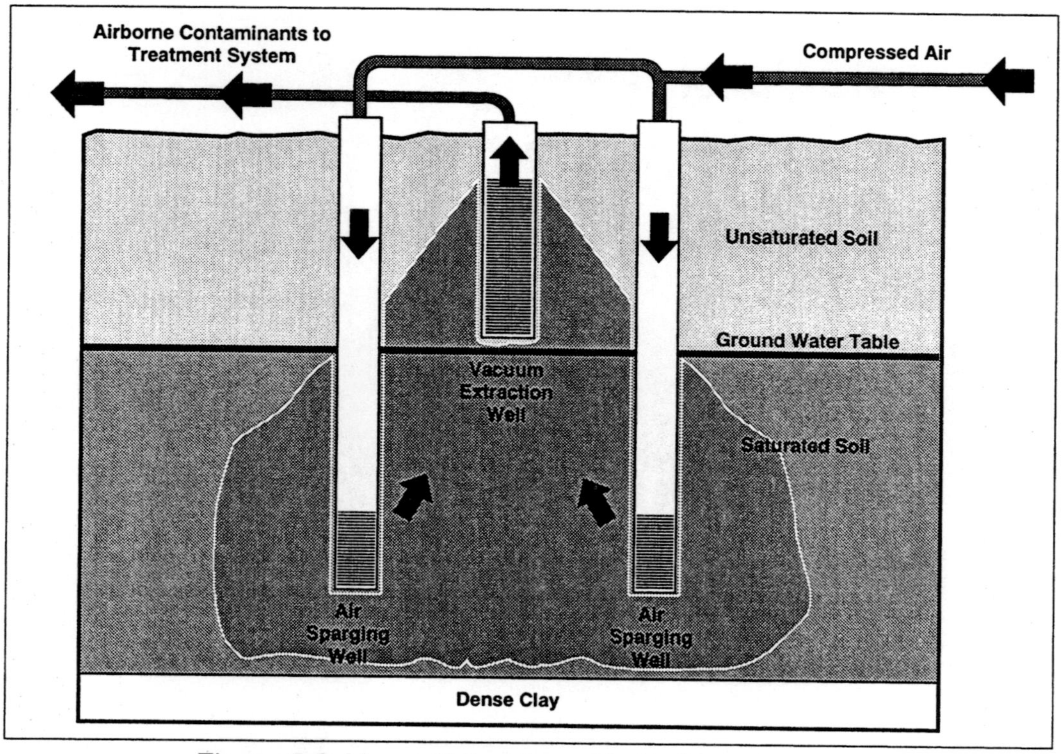

Figure 5-3. Vapor Extraction with Air Sparging

5.5 SURFACTANT FLUSHING AND CHEMICAL EXTRACTION

Chemicals generally coat the grains of impacted soils as a thin film on the grain surface. Impacted soils can be flushed with water mixed with a surfactant to leach the contaminants into the groundwater; the contaminant is then collected using downgradient wells. The method is termed surfactant flushing. A number of delivery and recovery methods are used. When other flushing agents are used, such as acids, bases, chelating agents, oxidizing agents, or reducing agents, the method is termed chemical extraction.

Some of the factors that favor removal of contaminants are:

- High water solubility of contaminants
- Low soil-water partition coefficient (a measure of the equilibrium between the soil organic content and water)

- Porous soil matrix

Techniques used for enhanced oil recovery (EOR) in the oil fields have been applied to in situ leaching, and a number of chemical additives have been considered. Texas Research Institute in bench and pilot scale studies modeled a sand aquifer. Multiple applications of percolation recovered up to about 80% of the gasoline, although the surfactant could not be separated from the recovered gasoline.

Implementation feasibility evaluation for full-scale site applications includes consideration of plume size, soil type, surface drainage and groundwater gradient. Delivery of surfactant can be done by forced or gravity methods. Gravity methods can work when soil permeability is greater than 10^{-3} cm/sec, and the depth of contaminated zone is less than 25 feet. These include ponding, flooding, surface spraying, and infiltration galleries. Forced method uses injection wells.

Equipment needs include mixing hopper, storage area for surfactant/water solution and chemical feed system. Recovery methods include gravity (open ditches, perforated pipe drains) or force (well points, deep wells, and vacuum wells). Flushing agents include Anionic (carboxylic acid salts), Cationic (Amines) and Nonionic (glycols) compounds. Treatment is necessary before the recovered water can be discharged to a POTW (Publicly Owned Treatment Works), usually under a NPDES (National Pollution Discharge Elimination System) permit.

Techniques for separation of surfactants from soil and process water, and better surfactant recovery are necessary before wider application. These methods are very effective in achieving cleanup of soils, but the recovered surfactants or flushing agents cannot be reused. As the methods involve introduction of another chemical into the subsurface, regulatory approval is not easy.

CHAPTER 6

EX-SITU REMEDIATION METHODS

Ex-situ remediation treats contaminated materials after removal from the subsurface. The actual treatment may be applied on-site or at an off-site treatment facility. A number of supplemental pretreatment and post-treatment processes are used to facilitate ex-situ remediation.

6.1 SUPPLEMENTARY PROCESSES

Pretreatment Methods

Crushing, Mixing, and Screening–All or some of these steps may be necessary to reduce the material to a suitable size or to separate the fraction that contains the contaminants. Methods employed are similar to those used for crushed rock handling at quarries and ore handling at mines. Appropriate setups of suitable processing equipment can be purchased or leased for on-site use.

Post-Treatment Methods

After the contaminants have been removed and treated, a matrix of clay, sand, sludge or solids is left behind. Significant quantities of contaminated water may also result from remediation. Air emissions, vapors or particulates such as dust may be generated.

Solids

The solids can be disposed of as fill or backfill using compaction, either directly or after stabilization. If the consistency, grain size and moisture are appropriate, and space is available, the solids can be used directly as compacted fill. If consistency or chemical content is inappropriate for direct use as fill, the solids can be stabilized by the addition of cement, lime or asphalt and then compacted as fill.

Water

Contaminated water containing dissolved chemicals may require treatment such as air stripping, carbon adsorption or biofiltration. Water containing particulates may need decanting or filtration. The pH of water to be disposed off may need to be adjusted before disposal for treatment at a POTW.

Air

Air containing particulates may require the use of baghouses or cyclones. Other post-treatment processes include use of carbon adsorption or catalytic oxidation for volatiles, and condensers or scrubbers for fine particulates and smoke.

6.2 EXCAVATION AND DISPOSAL

Excavation and offsite disposal of contaminated materials represents a complete elimination of the source. The equipment types generally used for excavation of soil include backhoes, cranes, dozers, and loaders. The excavated soil could be transported to an appropriate off-site disposal facility (landfill Class I, II, or III, or a recycling facility) based on the type and concentrations of contaminants.

For small quantities of shallow contamination, excavation is a simple, economical approach requiring no complex design. Mixtures of contaminants that may require multiple treatment techniques can be eliminated simply by using excavation and disposal. An excavation is relatively fast and it is easy to assess if all contaminants have been removed.

The depth to which excavations can be done economically is limited. Proximity to buildings also makes an excavation difficult and expensive. The wastes are not really eliminated or even treated, but simply transferred from one location to another, and the liability remains. If a large quantity of volatile compounds is brought up to the surface, there is a danger for exposure. As it is difficult to completely segregate the contaminated materials, the quantity disposed of is generally greater than the contaminated volume.

In the past, land disposal was the most common method for disposal of contaminated soils; however, certain wastes are now prohibited from land disposal, including certain solvents and dioxin-contaminated soils. Due to concerns about long term liability of landfill disposal, recycling of fuels and other petroleum impacted wastes is becoming increasingly common. The most commonly used recycling method for petroleum contaminated soils is thermal desorption. Reuse as road base after being blended with gravel or asphalt has also been increasing.

The handling techniques used in this remediation method are generally the same as for the excavation and grading of an ordinary uncontaminated site. Freshly excavated surfaces at contaminated sites may emit volatiles or dust particles that require special handling. Details of this remedial method are described below under five headings: Excavation Planning, Dewatering, Shoring and Underpinning, Grading, and Handling of Particulate and Volatile Emissions.

Excavation Planning

Excavation and offsite disposal is feasible only if the contaminated material can be accessed from an open area. During excavation planning, adjacent structures and utilities requiring underpinning or shoring are identified. Even though environmental permits and agency approvals for remediation may be in place, excavation and grading require additional permits from the grading departments of the local building and safety jurisdictions, and the local air quality oversight agency. It may be prudent to do a crack monitoring survey and a general building condition survey of the existing adjacent facilities to avoid future damage claims. As a minimum, exterior photo-documentation of the adjacent fences, walls, and buildings should be performed.

Areas for stockpiling and segregation of excavated materials should be identified in advance. Ideally, all areas for stockpiling, loading, and special storage for contaminated water should be staged on-site. However, if the whole site is to be excavated and if adjacent land is available, arrangements can be made to rent or lease the land for storage, laydown areas, or truck staging areas to minimize disruption to excavation operations and to increase efficiency. Consideration should be given to community relations, traffic control, haul routes for transportation of material for disposal, and import material sources.

Excavation

The excavation of contaminated materials is the first step. Excavation methods are dependent on the nature and consistency of the matrix in which a chemical of concern is contained. Dry sand, wet clay, silty sludge or peat all require slightly different equipment and methods. Many of the techniques are adapted from earthwork, grading and solid waste handling operations. Deep excavations in urban areas may require shoring of sidewalls, underpinning of adjacent buildings and support for utilities in conjunction with careful planning and monitoring. Other constraints that may be imposed on any excavation in urban areas include: limited working hours, traffic plans and controls, control of fugitive dust and other emissions, community relations, and public awareness programs. Permitting issues sometimes predominate over technical issues.

Dewatering

At sites where excavation is planned to be deep enough to approach or go below the water table, it will be necessary to dewater the site by predrainage. This improves trafficability for the excavation equipment. At a site where excavation depth is shallower than the water table, water from seepage, possible perched groundwater, precipitation, or leakage from utilities will need to be managed. For such incidental occurrences of water, it is practical to handle the water as the excavation is proceeding. Surface infiltration must be controlled by constructing berms at the perimeters of the excavation.

Predrainage for lowering the water table requires an assessment of the water bearing layer by means of pump tests before the dewatering system can be designed. Once the aquifer properties and pumping rates are known, the number of wells around the site and placement of pumps can be planned (Figures 6-1 and 6-2). Sufficient advance time is required, as some sites may need several weeks of pumping before the water table is lowered adequately. Dewatering reduces the water present in the contaminated materials, thus reducing the weight that may need to be hauled, dried, treated or buried. The requirements for dewatering vary at each site, and artificial barriers such as sheet piles or slurry walls may also be used to reduce or direct groundwater away from the excavation.

NPDES permit is required in addition to permits from the local municipality to discharge the dewatering water from a contaminated site into the nearest storm drains (clean discharge) or sewers (mildly contaminated). Heavily contaminated water requires on-site treatment, or hauling away by means of tanker trucks to recycling facilities. Either option is very expensive. Testing of effluent must be planned, both as a permit requirement and as a prudent practice for avoiding claims of pollution discharge.

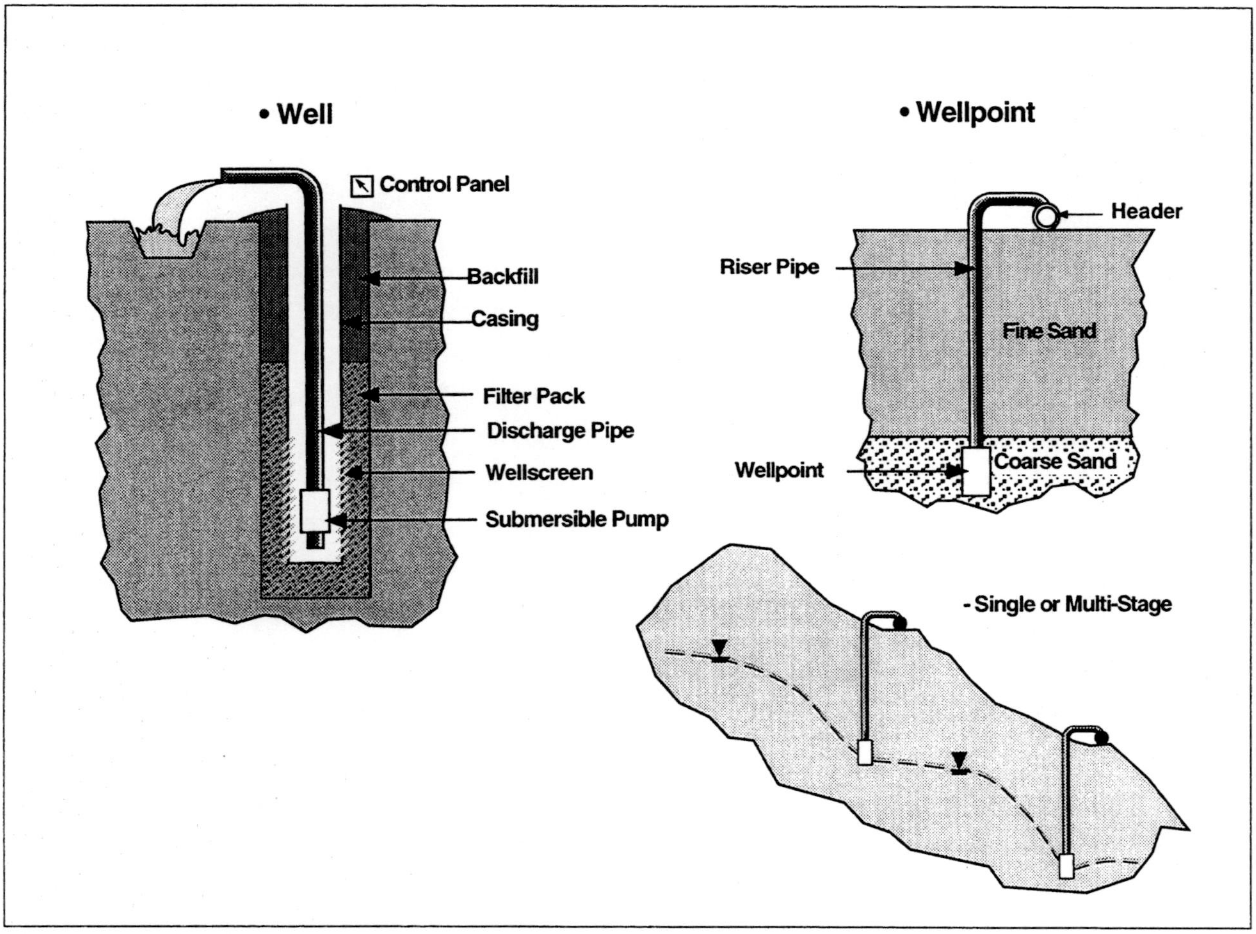

Figure 6-1. Dewatering- Wells and Wellpoints

Condition	Deep Wells	Wellpoint System
Soil		
Silty/Clayey Sand	Poor to Fair	Good
Clean Sand/Gravel	Good	Good
Schedule		
Rapid Dewatering Required	Unsatisfactory	Ok
Slow Dewatering Permissible	Ok	Ok
Excavation		
Shallow (<20ft.)	Ok	Ok
Deep (>20ft.)	Ok	Need Multistage
Capacity		
Per Unit	25-3,000 gpm	0.1-25 gpm
Total System	200-60,000 gpm	Low-5,000 gpm

Figure 6-2. Dewatering- Selection of Pre-Drainage Methods

Shoring and Underpinning

A stable pit can be excavated by merely sloping back the sides. However, it may not be possible to slope back if adjacent buildings, roads, utilities or other facilities are in the way. In such cases, it is necessary to shore the excavation sides by installing structural elements. Depending upon the depth of the excavation, the earth pressures, and the settlement sensitivity of the adjacent improvements, shoring can range from simple timber sheeting to tied-back soldier piles (Figure 6-3). Shoring installation increases the expense of an excavation significantly, although it will reduce the volume to be excavated. In most cases, the safety against slope failures is also increased. Most shoring installations require periodic monitoring to ensure compliance with deflection criteria.

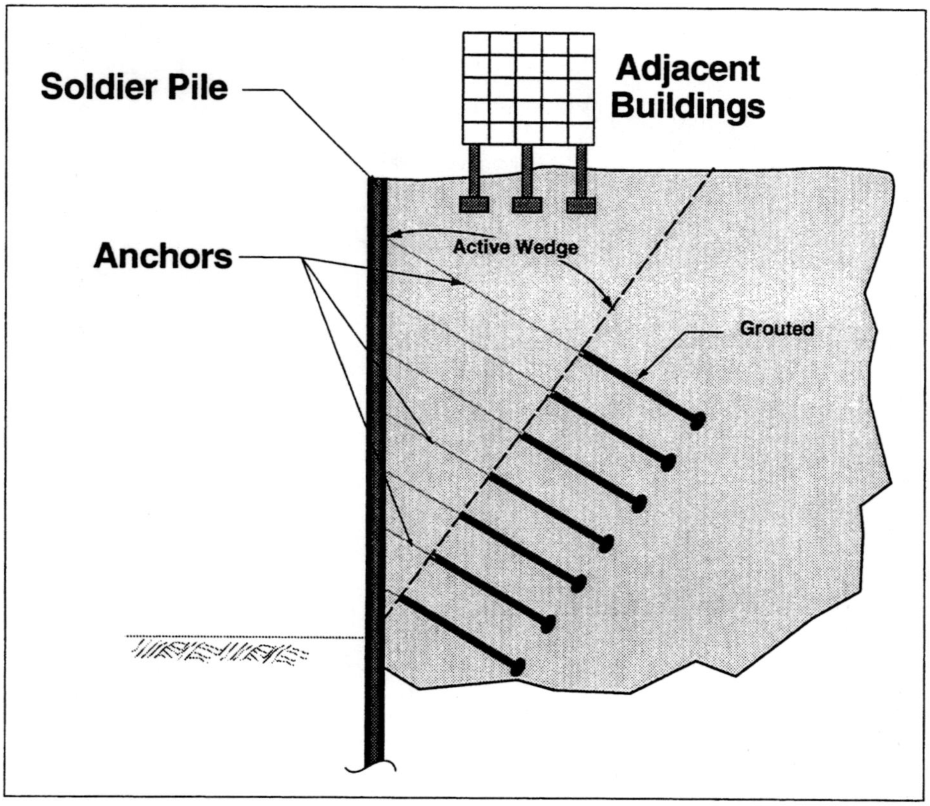

Figure 6-3. Shoring Tied-Back Excavations

Underpinning of adjacent structures to prevent settlements during excavation requires an understanding of their foundation systems, and design and installation of supplemental foundation supports such as piles or spread footings to achieve structural integrity and stability during excavation. If the existing foundation system is robust enough, it may be sufficient to add strengthening elements such as grade beams. For simple inexpensive improvements such as concrete block walls or fences, it may be more economical to take them down and rebuild later, rather than going through the expense of designing, installing and monitoring a shoring system.

Grading

Grading refers to both the removal of contaminated materials, and backfilling with clean soils. Once a site has been dewatered and shored, the actual excavation can begin. The preplanned staging areas can be used for stockpiling, segregation, and loading of materials to be hauled away to the disposal sites. Stockpiling may be necessary to determine the average chemical concentrations, and thus decide the appropriate disposal locations. Bulking or swell upon excavation increases the volume of material that needs stockpiling and hauling (Figure 6-4).

Appropriate transportation manifests accompanied by test results will be required by the disposal facilities before accepting the excavated material. Some of the material may be sent to recycling facilities. A project diary, compilation of test results and photo documentation of excavation and stockpiling operations should be maintained.

At the end of excavation, confirmation sampling and testing of side walls and excavation bottom is done to confirm that all contamination has been removed. Backfilling can then begin using clean on-site material or imports. If some of the contaminated material is to be left unexcavated, a synthetic liner may be used as a barrier. To satisfy compaction requirements specified in the contract, laboratory testing will be needed to determine the reference values in accordance with ASTM (Modified Proctor) test methods. Records of test locations, depths, and percent compaction results are necessary for obtaining final approval from the grading departments that backfilling of the site has achieved compliance.

Emissions

Compliance with the site permit requirements from the local air quality oversight agency must be maintained. Both particulate and volatile emissions have to be monitored and controlled during most remediation projects. During excavations large surface areas are exposed, requiring very intense compliance controls. Controls such as water spray, use of vapor suppressant foam, covering with tarp or plastic covering, and suspension of activities on windy days may be necessary to achieve compliance. Nuisance complaints by site neighbors to the local agencies must be handled promptly.

Monitoring data often become the basis of regulatory oversight by agencies. There may be ceilings or limits for the total emissions allowed from an excavation project, so cumulative totals must be kept. Upwind and downwind stations for monitoring of total suspended particulates, and PM-10 particulates (finer than 10 microns–the breathable dust) will be necessary, as well as a meteorological station for monitoring of wind speed and direction.

The excavation may need to be suspended if wind speed exceeds a certain limit, or if the ambient air quality is poor due to atmospheric conditions. Alternative excavation faces can be opened in less contaminated areas of the site on certain hot days to keep the volatiles under check. Hand-held monitoring instruments will be necessary for worker safety and to guide the excavation operations.

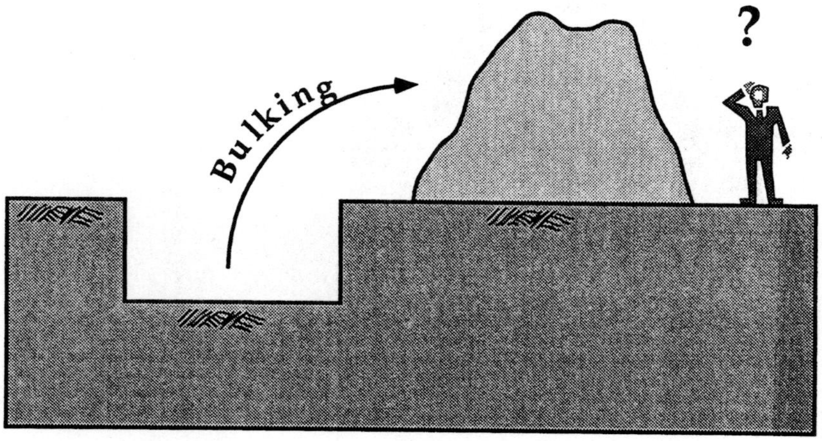

SOIL TYPE	MOISTURE	FACTOR
Gravel / Sand	Wet or Dry	5%
Silt / Fine Sand	Wet	15%
Clay with Sand / Silt	Wet or Dry	30%
Organic Soils	Dry	45%

Figure 6-4. Bulking or Swell

Case History: Construction at a Contaminated Site

A manufactured gas plant site located in Venice, California, where contamination had reached the ground water (Figure 4-2, discussed earlier), was slated for redevelopment. An advertising agency selected the site for its headquarters building, and a design was prepared by the world–renowned architect, Frank Gehry. As construction began, contamination was discovered, the site neighbors complained of odors, and the construction was halted. Characterization showed that contaminates included heavy hydrocarbons and PAH's (Polycyclic Aromatic Hydrocarbons), including Anthracene and Benzo (a) Pyrene. Few volatile organic compounds were detected, and PAH's were considered the main health risk. The PAH's are carcinogenic, and the property was declared a State Superfund site.

The building plans included a three-level subterranean parking garage, necessitating a 25-ft deep excavation. The contaminants were not very mobile and did not have to be removed. However, the building required an underground garage, and various options were studied for the excavation. Further testing of toxicity showed that because of low mobility the PAH- containing soil could be disposed off at ordinary municipal landfills. The main challenge was to safely excavate and transport the soil. The main hazard during the excavation was due to the dust that could escape the site, and a total dust "ceiling" was negotiated based on health risk analyses. A detailed excavation plan was prepared after public hearings and input.

The subterranean portion of the building also had to be protected against the contaminates that were going to be left behind under the surrounding properties and streets, all of which were part of the manufactured gas plant at one time. An 80-mil High-Density Polyethylene (HDPE) liner was selected as the protective material for the underground portion.

The excavation operations are illustrated in Figure 6-5. An 8-ft high wooden fence that had an entry gate and an exit gate protected the site; both were controlled gates with flagmen. A truck queuing station allowed only two trucks at a time to pass, the remainder of the trucks waited at a holding area about 1-½ miles away, and were radio dispatched. Fresh plastic liner was placed in the truck beds as they waited outside the gate, after which the trucks proceeded inside the site.

Six dewatering wells operating for about six weeks before excavation began had predrained the site. Collectively the wells discharged about 1.1 million gallons per day to the storm drains under an NPDES permit, which included requirements for regular testing of the effluent. The excavated soil was stockpiled in an area near the entry gate where the trucks were loaded by a front-end loader. A foaming and watering station near the fresh excavation face was used on an as-needed basis to control dust and emissions emanating from the soil. The excavator operator carried a full-face respirator that he donned each time the vapor concentration exceeded the guidelines. Continuous monitoring of vapors at the excavation face was done by an operator using an OVA on a full-time basis.

After loading, the truck proceeded to the decontamination station near the exit gate. A second station for foaming and watering was also available. The soils in the trucks were watered or covered with vapor suppressant foam depending upon the measured emissions, and weather conditions. The intent was to keep the emissions to a minimum on the way to the landfill. The truck was also covered with a tarp and secured at this station before exiting the site. The upwind and downwind particulate load was measured by using two samplers, one for the Total Suspended Particulates and the other a PM-10 sampler for particles finer than 10 microns, or the breathable dust. The difference between the upwind and the downwind loads was considered the amount of dust that escaped the site during the workday. As the site was only about three-quarter of a mile from the beach, the prevalent wind direction reversed during the day. At daybreak the breeze was coming from land as the seawater was at a higher temperature. *(continued on next page)*

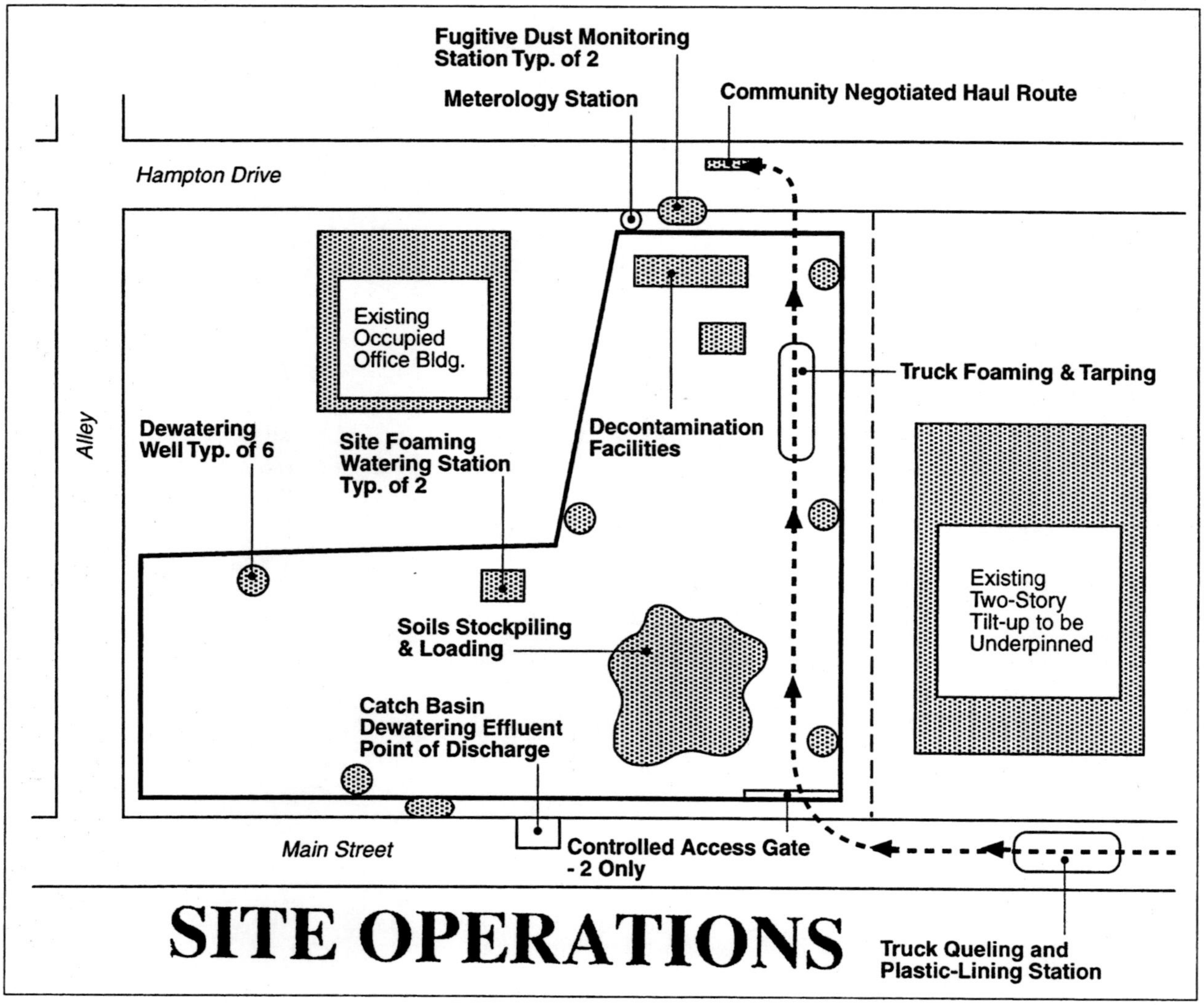

Figure 6-5

By about 10:30 AM the land surface heated up and the wind direction reversed. At this time the filters were reversed between the upwind and the downwind stations. In this way the upwind and the downwind filters functioned as such all day, even though the wind direction reversed.

The net amount of dust that escaped each day was measured, tracked as a cumulative total and compared with the project ceiling of 1.38 grams of dust/liter of air. In addition, there was also a daily dust limit of 400 micrograms/liter. Other regulations regarding dust escaping a construction site were also observed. The excavation was successfully completed within the project ceiling for total dust emissions. After the excavation was completed, an HDPE liner was placed, the subterranean garage completed and the building constructed.

6.3 BIOREMEDIATION

The Process

Bioremediation is a treatment process that uses microorganisms to decompose and transform toxic or hazardous contaminants into nonhazardous compounds. In the case of petroleum hydrocarbons, the end products are carbon dioxide and water. Bioremediation uses naturally occurring or genetically engineered microorganisms. The process decreases the level of toxicity in the organic materials and reduces the migration potential of hazardous components in the media being treated, including soil, water, sludges, and industrial residues.

The use of microbes in the cleanup of toxic waste has been practiced for several decades in treating municipal wastewater and sludges, refinery waste, and other chemical waste streams.

Applicability

Bioremediation has been found to be suitable for a wide range of petroleum contaminants. As a remedial technology, it has a low operating cost and does not require much maintenance. As the wastes are converted into harmless products, the long-term liability is reduced. it is the ultimate "natural" method for remediation and is considered ecologically sensible. If space is available and the setting is industrial, as at many brownfield sites, remediation can be performed right on the site without mobilizing any expensive plant and equipment. As microbes are always active, remediation occurs literally 24 hours a day. The process can be coupled with other techniques, depending on the type of contaminants present.

As with other natural, organic processes, bioremediation takes considerable time and cannot be accelerated by much. The applicability is excellent for fuel hydrocarbons and petroleum, but limited to nil for chlorinated compounds or metals. Although it is rare, but occasionally microbial degradation may lead to more toxic or mobile products than the parent compounds. The basic process is self-sustaining, but environmental conditions need frequent adjustments for optimization, and the process has to be adapted to each site and the environment.

Microbial Requirements

Microbes utilize organic compounds including pollutants as an energy source for their own metabolic process. For this process to take place, a favorable environment must be maintained. Although there is no one set of environmental factors ideal for all organisms, microbes are active within certain ranges of temperature. The most suitable temperature for mesophiles (which are used for petroleum hydrocarbons) is between 25 to 35°C. Soil moisture is also important since the microbes transfer nutrients and wastes across their cell walls in aqueous system. The media moisture should not exceed the level that interferes with aeration or oxygen diffusion, but should be above that of wilting point for plants (0.3 bars) as the lower limit. Soil aeration is necessary for maintenance of aerobic conditions for microbial growth. Another factor that aids bioremediation is pH of the soil. The optimum range of pH for most soil microorganisms is between 5.5 to 8. Microorganisms also need the first 28 elements in the periodic table in trace amounts. Usually there are sufficient quantities of these elements present in the soil, but nitrogen and phosphorous have to be supplemented.

Landfarming

Landfarming is the oldest bioremediation technique. It has been in used since prehistoric times for managing farm wastes. With the arrival of the industrial age, came industrial wastes. The same disposal techniques were applied with mixed results. In landfarming, the growth of bacteria is enhanced by creating and maintaining conditions that promote microbial growth: moisture, air, growth medium and temperature. Aerobic bacteria require moisture for growth, but also need a well-aerated inert matrix for oxygen supply.

The actual application of landfarming is limited to heavy hydrocarbons, as the lighter-end volatiles are likely to escape before they can be digested by the bacteria. The contaminated material is mixed with an inert medium such as sawdust or fine gravel and laid out in windrows. It is then periodically aerated by turning over and kept moist using water spray. Essential nutrients such as phosphorus and nitrogen are added as fertilizer. Periodic testing of the windrowed material is necessary to keep track of the rate of remediation, and to know when the concentrations have been lowered to the target levels necessary for compliance. From a few weeks to several months are needed for this method of remediation.

For bioremediation to be effective, factors such as chemical and physical properties of contaminants (hydrocarbon concentration distribution, metals content), population density of microorganisms, soil physical and chemical properties (pH, particle size, organic carbon, nutrient availability, etc.) must be maintained.

Temperature influences the metabolic activity of microorganisms because of their enzyme reactions. The optimum temperature for land farming ranges from 18° to 30°C. Minimum rates would be expected at 5°C.

The soil moisture content of the contaminated soil affects land farming due to increased dissolution of the residual compounds and dispersive action, and fulfills the need for microbial protoplasm. Either extreme, water logging or desiccation, will reduce effectiveness of the operation and the biodegrative activity. The soil should be maintained at approximately 50 to 70% of its water-holding capacity.

Land farming is reported to be most effective when the soil pH is maintained at about 6.5 which is the pH at which microbial activities will not be limited. Maintaining soil pH at 6.5 also enhances the immobilization of certain waste constituents.

Biocells

Not all sites have the luxury of space, air quality laws, and suitable climate that would permit open air land farming. Another on-site method of remediation is by using biocells (Figure 6-8). Contaminated material is mixed with soil amendments to provide the right texture, and then stockpiled in soil piles over pipes that supply water and nutrients. The whole assembly is then covered under thick, durable plastic such as high density polyethylene (HDPE), and sealed. In addition to monitoring for contaminant level, moisture content and nutrient content, the biocells require monitoring for rodent and insect control.

6.4 SOIL WASHING AND RELATED METHODS

Soil washing is a method of remediation in which excavated soil is washed by using water mixed with surfactants. The washing or flushing action is similar to that employed for the in-situ treatment method termed Surfactant Flushing and Chemical Extraction. However, as the soil has been separated from any subsurface moisture or groundwater, the volume of liquid that requires handling is much smaller and much more manageable. The contaminated soil to be washed is deposited into a large tank where the washing process is facilitated by the use of mechanical agitators. The temperature, holding time, rate of agitation, and type of surfactant can all be varied easily for highest effectiveness. The following paragraphs describe the soil washing process, followed by the two related methods: Chemical Extraction and Chemical Treatment.

Soil Washing

The soil washing process can be divided into six steps described in separate paragraphs below:

Pretreatment

This step involves excavating the contaminated materials and preparing them for soil washing. Any debris, large items, cobbles, pieces of underground improvements, concrete footings or pipes, are separated. The soil is stockpiled into manageable sized batches for handling.

Separation of Soil Fractions

Sandy and gravelly materials are separated from the finer clay and silt fractions. If the material is dry enough, the separation is accomplished by use of screens and sieves. If some clumps or clods have formed, crushing facilitates the separation process. If, however, the soil matrix or the contaminant chemical are such that dry separation cannot be achieved, wet separation is needed.

For wet separation, batches of material are placed in tanks and agitated. The coarse materials—sand sized and larger particles—are allowed to settle, and the finer fractions decanted as a suspension. This step may need to be repeated a number of times until most of the coarse grains have been separated. Most of the contaminants are adsorped on the fines, saving the need to treat the coarse grains.

Treatment of Coarse Fraction

The sand-sized and larger particles have relatively limited surface area, and do not contain much adsorbed or absorbed chemicals. Washing of this fraction is accomplished relatively effortlessly and the soil matrix should test clean after a short holding interval in the agitator tank. The fraction is then allowed to drain and dry out. After drying, it may be clean enough to be used as backfill. At worst, this material is disposed of at a municipal landfill at same rates as daily cover or ordinary refuse.

Treatment of Fine-Grained Fraction

The fine-grained fraction includes clay and silt sized particles that from a suspension during soil washing. Particles in this size range represent most of the surface area and contain much of the adsorbed contaminant. Treatment for this fraction represents the actual treatment of the contaminant. Typically, it is first attempted to see if the use of surfactants or other agents can induce the contaminants to desorb from the grain surfaces. Mechanical processes that might aid this effort include soaking, agitation, stirring, heating, decanting and filtering. Even at the end of all efforts, it may become uneconomical to try and remove all of the chemical contaminants from the finer fractions. The remainder is disposed off as sludge after draining and removing as much of the water as possible.

Treatment of Process Water

During each of the above steps, supernatant water or liquid is separated, and contains most of the contaminants, washing agents and surfactants. This water most likely will be too polluted to be discharged into a sewer line leading to a POTW (Publicly Owned Treatment Works),thus requiring treatment before discharge. The treatment may simply be adjusting the pH to bring it to within discharge criteria, or a specifically designed industrial treatment process. In each case, a NPDES permit will be needed, and there will be an interest in reducing the volume of the discharge water to a minimum. If the contaminants were nonvolatile, it may be possible to reduce the volume of process water by simple evaporation, especially if the discharge is metered for disposal. In the case of fuel grade contaminants with sufficient octane value, recyclers may be interested in receiving the process water.

Residue Management

During the various treatment steps such as filtration, decanting, etc., residues are left over containing various chemical fractions and grain sizes. The residues can be added to the sludge from the fine-grained fraction. If either the grain size of a particular residue or its chemical content is unsuitable for such inclusion, it will be necessary to dispose of the residues separately.

Chemical Extraction

If simple washing using surfactants or detergents is considered unfeasible, or proves to be unworkable during bench-scale testing, then stronger agents such as solvents may be needed. Chemical extraction equipment setups resemble chemical manufacturing plants. Because of the presence of solvents the use of open tanks or holding areas is not appropriate. Also, because of the cost of the agents involved, there may be greater effort spent in recovering and reusing them.

Chemical Treatment

If even the use of solvent grade reagents is not effective in separating the contaminants from the inert matrix, then the nature of the adsorbed chemicals needs to be chemically altered. The intent is to either destroy the contaminants or at least change them to a more benign form, making the disposal easier and less expensive. The details of the processes are available from vendors; as most of chemical treatment methods are patented processes that employ proprietary agents. It is difficult to independently assess the process validity, or to evaluate the vendor's success claims.

6.5 THERMAL DESORPTION

This technique uses heat to separate the contaminants from inert media. Treatment is useful for VOCs and Semi VOCs. Temperatures used (100-500°F) are lower than temperatures used during incineration (over 1000°F) and time-at-temperature is shorter; consequently, the regulatory requirements are less stringent. The four components of thermal desorption are: Pretreatment, Thermal Desorption, Gas Post Treatment, and Solids Post Treatment.

Pretreatment

Excavation of contaminated materials is the first step for thermal desorption. After excavation, the material is transported to the proximity of the desorption equipment and stockpiled. Storage of the excavated material allows staging and gradual feeding of the material into the desorber. Oversized debris is removed, and any particles of very large diameter are separated. Any lumps or clods can be crushed to make the feeding of material into the desorber easier. If moisture content is high, the water is allowed to drain away.

In the case of certain wastes, there is a need to neutralize the contaminated material to limit the formation of acid gases during heating. The final step in pretreatment is to prepare the feeding system, consisting of either a conveyor belt or a loader, for delivering the contaminated soil to the feed hopper of a desorber.

Thermal Desorber

The desorber generally includes a feed hopper into which the contaminated material is fed, and a screen gate to remove any cobbles or debris. The feed hopper leads into a drum or kiln—a cylindrical container— in which the material is heated so the adsorbed contaminants can volatilize, leaving behind the inert mineral matter. Most desorber units accept continuous feed, allowing a certain heating time for the material in the drum before exiting. The heat source is generally a natural gas or propane burner, providing easy temperature control to compensate for varying moisture content and grain size. After adequate heating, the solids are discharged out of the drum. Desorbers are indirectly fired, so the flame does not come in contact with the contaminated materials. A direct-fired kiln where the flame is allowed to touch the contaminants would be considered an incinerator.

Gas Post Treatment

The off gases coming out of the desorber drum are fed into a heater or a catalytic oxidizer in order to burn off any hydrocarbons. The temperature of off gases rises significantly during desorption and oxidation stages, and they are then led into an air exchanger. The air exchanger is a cooling device or an air radiator that cools off the off gases before their entry into the baghouse, the final device used in gas post treatment. This prevents any singeing or other heat damage to the baghouse, where off-gases shed any particulates. The baghouse is a large canvas cylinder in the form of a bag similar to a vacuum cleaner. This prevents the off gases from introducing excessive dust into the atmosphere.

Solids Post Treatment

The treated solids exiting a desorber are poured into a pugmill to break down any clumps that may have formed and thus bring the particle size into an optimum range. The crushed materials are in a very hot and dusty condition, and require the use of a rehydrator, which may simply be a water spray system to increase the moisture content of the treated soil. This prevents excessive dust from escaping into the atmosphere. After the rehydrator, the treated solid material is transported back to the site from which it was excavated, used as fill at other sites, or disposed off as daily cover material at municipal landfills.

6.6 STABILIZATION

Techniques

Stabilization refers to a group of methods aimed at reducing the mobility of contaminants and improving the handling characteristics of the resulting material. This improved material is then compacted into a stable mass.

However, a thorough understanding of the potential behavior of stabilized and solidified waste is necessary to make judgement about the long term effectiveness of the containment. Stabilization techniques include fixation, encapsulation, and solidification.

Fixation is a chemical process that fully or partially bonds the waste by adding a supporting medium, binder, or modifier to the soil.

Encapsulation is a physical process by which a solidification agent surrounds waste particles and binds them together in a mass.

Solidification is defined as the process by which a liquid or semiliquid sludge is mixed with a sufficient quantity of solids to allow landfilling of the mixture with conventional landfilling equipment, regardless of whether fixation or encapsulation has occurred.

All three techniques are generally simple, can be implemented with readily available equipment, and are generally less expensive than land disposal for hazardous wastes. There is a relatively high throughout rate resulting in shorter treatment times compared to many other technologies. Organic content of wastes often interferes with the stabilization process. Once the waste has been stabilized, the resulting product may still require handling as a hazardous waste.

Stabilization increases the average particle size, thus reducing the potential for particulate emissions. The overall goal of all these techniques is either the safe disposal or reuse of the waste material and is achieved by one or more of the following:

- Improving the handling and physical characteristic of the waste.
- Limiting the solubility rate of the compounds in the waste.
- Reducing the permeability of the soil matrix.
- Reducing the surface area of the waste in contact with solute when leaching occurs.
- Reducing or eliminating the toxicity.

The US EPA has recommended applications for which stabilization techniques are effective. Applications include treatment of soils and sludges containing metals, asbestos, radioactive materials, inorganic corrosives and cyanides. However, these techniques are considered ineffective for treating halogenated and non-halogenated volatiles, and only partially effective for treating PCBs, pesticides, dioxins, furans, and organic cyanides and corrosives. Concern about the long term integrity and leachability of chemically stabilized wastes has delayed the acceptance of these techniques as a disposal option for hazardous waste.

CHAPTER 7

INNOVATIVE AND INTEGRATED REMEDIAL TECHNIQUES

7.1 INNOVATIVE TECHNIQUES

Two innovative techniques that involve heavy use of electricity have been demonstrated up to the pilot stage in studies sponsored by the Electric Power Research Institute (EPRI). However, neither has come into common use. Both are described below.

In-Situ Vitrification

Soils are converted into durable glass by melting in-place using electricity, and in the process, volatilizing some contaminants and hardening others along with the vitrified soil. Originally, the technology was used to stabilize high level radioactive wastes in canisters or drums. When applying in-situ, four electrodes are inserted into the ground 3 to 5 meters apart. A mixture of graphite and glass is placed in an X pattern for conduction. As current is passed through electrodes, internal resistance of the soil causes temperature rise, which results in melting of glass and other materials. As soil melts, it becomes conductive, allowing outward spreading of the molten zone. The electric current must then be increased to as much as 3,000 to 4,000 KW. Less dense materials such as organics rise to the surface. An off-gas hood is included as part of the system to collect volatilized chemicals, which are passed on to a treatment area. After stoppage of electric current, one to two weeks are required to cool off, as the molten zone often rises to temperatures over 1,700°C (silica and aluminum oxides melt at temperatures between 1,100 to 1,600°C).

A case study by EPRI on in-situ vitrification of PCB in a pilot test showed 0.05% release into the off gas system, 0.7 mg/Kg left over in surrounding soils and 99.99% efficiency for vitrification.

Costs are likely to be $150 to $220 per cubic yard. Longitudinal metallic objects such as pipes or bars can short out electrodes, but do not make a difference for up to 70% of the distance.

Off gases require treatment, and are the primary exposure source. Contaminants are combusted or destroyed. The technology is extremely energy intensive and expensive.

In-Situ Microwave Volatilization

The installation for a microwave volatilization set-up is similar to in-situ vitrification. A source of electricity is brought to the site and a target is selected. The microwave generator is installed in a pit or shaft below the ground surface and connected to the source of electricity. The range is limited, and the application is aimed at contaminants that can be volatilized.

Both in-situ processes involving use of electricity are not very common. Inhibiting factors include high cost of electricity, possible hazards of high voltage and currents, and potentially uncontrolled emissions.

UVB™ Extraction

The UVB-Technology (German: Unterdruck-Verdampfer-Brunnen, patented by IEG-mbH, D-7400 Reutlingen) has been utilized in Germany for many years for the in-situ remediation of groundwater contaminated with volatile pollutants. The UVB Technology is especially suitable for eliminating biologically degradable contaminants (e.g., hydrocarbons, phenols, pesticides, etc.) from the groundwater without having to pump the groundwater to the ground surface.

The specially designed UVB cell produces a groundwater convection cell in the aquifer around the remediation well. The circulating groundwater constantly transports both contaminants and existing degrading bacteria to the well. When flowing through the well the contaminants are adsorbed onto the material inside the bio-reactor; simultaneously, the bacteria settle in the same area. If necessary, the accumulated microorganisms can be supplied with nutrients.

Another advantage of the UVB is the oxygen enriched groundwater, which enhances the population growth of the microorganisms in the aquifer, thus accelerating the degradation process. System variations include discontinuous circulation flow, reversing the circulation direction, installation of different bioreactor configurations and materials, and using a combination of physical, chemical and/or biological methods. These variations enable the technology to easily adapt to different contamination sites.

Other Methods

A number of other innovative techniques are being tried:

- Electro-Acoustic Soil Decontamination
- Electroosmosis
- Bio-Recovery System for removal of metal ions

- Reverse Osmosis
- Soil Biofilter
- Thermal-Plasma Arc
- Bioremediation using carbon/oxygen isotopes
- Removal of hazardous organic pollutants by Biomass Adsorption
- Resin Adsorption
- Ultrafiltration

7.2 INTEGRATED TECHNIQUES

Contaminated sites that contain a complex mix of chemicals are often resistant to treatment by using a single method of remediation. At these sites it is advantageous to integrate two well-proven techniques, by using them in tandem or in series. The techniques are integrated either to attack distinct materials, or to enhance the effectiveness of the primary method of treatment. For instance metals and volatile compounds may require a desorption technique, followed by a stabilization treatment. Some examples are listed below.

- Thermal desorption followed by cement stabilization (focusing on two dissimilar contaminants)
- Groundwater pump and treat with reinjection (enhancing the primary treatment method, Figure 7-1)
- Vapor extraction with bioventing (enhancement)
- Pump and treat with ultra-violet treatment (augmentation)

SITE Program

The U.S. Environmental Protection Agency's Superfund Innovative Technology Evaluation (SITE) Program serves several purposes, including (1) the development and implementation of innovative treatment technologies for hazardous waste remediation and (2) the development and implementation of monitoring and measurement technologies for evaluating the nature and extent of hazardous waste site contamination. The SITE program is administered by EPA's Office of Research and Development and consists of four interrelated programs.

- Demonstration Program
- Emerging Technologies Program
- Monitoring and Measurement Technologies Program
- Technology Transfer Program

Further details on this program, including opportunities for participation, are available through the EPA.

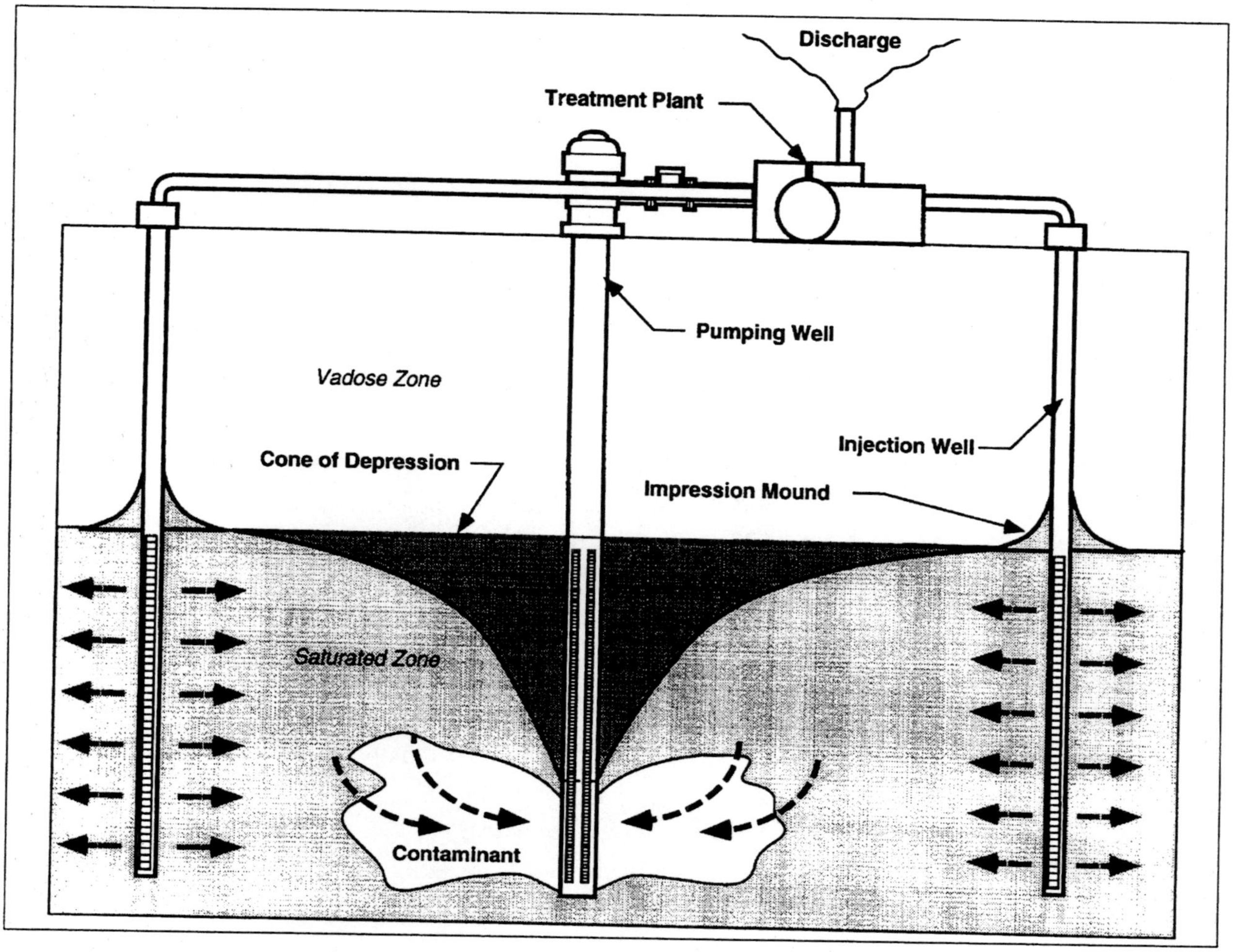

Figure 7-1. Groundwater Pump-and-Treat with Reinjection

PART III

SITE CLOSURE

CHAPTER 8

ENVIRONMENTAL DATA MANAGEMENT: INFORMATION TECHNOLOGY APPLICATIONS

8.1 INFORMATION ARCHITECTURE

This chapter reviews the current trends in environmental data management. This field is changing very rapidly and the available tools are increasing in complexity and computing power. Although this chapter can point the way towards new developments, many of the currently available software and hardware capabilities are likely to be exceeded before very long.

The amount and complexity of environmental data can be so large as to defy intuitive understanding or orderly organization. At the same time, the need for timely access, and the financial or legal implications of inaccurate or incomplete information can be significant. State of the art information management systems are being developed to handle complex data and to aid the decision makers. With the advent of powerful desktop computers and Internet-based or intranet-based access, the availability of such decision-making tools has increased.

An important initial step is to decide upon the uses for the environmental data. Based on use, a suitable information architecture can be devised. An appropriately designed system will always allow access to the specific data and images required at any particular moment. An expertly designed system fed by streams of collected data with easy to use interfaces, potent analysis capabilities, and an assortment of means of communication can be a powerful tool.

The most efficient method for designing a system is with the use of customized modules. Such modules can be deployed quickly and cost effectively. Network technologies can be exploited to increase access and reduce cost for multiple users. Typically, a versatile system will include analytical or modeling tools, databases such as GIS, and visualization tools. Often, the analytical portions are handled by engineering or geosciences specialists, and the output from their work is imported into the programs operated by information technology professionals.

Analytical and modeling tools are too complex and specialized to discuss here, but the other two components, databases and visualization programs, are discussed below.

8.2 GIS AND OTHER DATABASE PROGRAMS

GIS—Geographic Information Systems—are databases that store information with reference to the geographic location from which each data item originated. Thus, the results of analytical tests performed on soil samples obtained from borings are stored with reference to the geographic location of each boring and the depth from which the samples were obtained. Relational database schemes are used to store the information. A lot of the data being acquired these days is space-dependent data. Particularly useful is the storage of continuous data from variable media such as groundwater, allowing interpolation and extrapolation of the data in an effective manner. Specific software such as Arc-Info™ or Intergraph™ provide powerful options for data entry, data search, and display.

Other database programs allow the use of similar or less sophisticated and simpler schemes for arranging, storing and retrieving the information.

8.3 VISUALIZATION TOOLS

The data resulting from environmental investigations is often complex and voluminous. Professionals who work with it daily often require hours or days to digest it adequately and visualize the spread of contaminants. Others who have not undergone formal training in geology, chemistry or hydrology may never fully grasp the significance of the information. When presenting a remediation proposal to a lay audience or to specialists from a different discipline, it is helpful to illustrate the information for easier understanding. A number of different visualization tools are used for this purpose.

Aside from data illustration, another use of illustrative and visualization tools is to present the result of computer modeling in an understandable manner. A number of sophisticated numerical models are available for interpretation and prediction of the behavior of contaminants. Predicting the flow of contaminated groundwater in the subsurface requires the use of some of the most complex programs. This area also offers the most challenge in trying to provide an overarching explanation for isolated data points from groundwater monitoring wells observed over long periods. Even though the modeling programs are mathematically sophisticated, and their results impressively prescient, their output consists merely of tables full of numbers. Visualization tools are often the answer.

The simplest step in visualization is to take the results form an analytical program and import them into an illustrative program. At one former gasoline station site vapor extraction was selected after successful pilot testing. However, the site became the subject of a lawsuit, and it became necessary to prepare illustrative materials for use in the court, if necessary. The results of soil borings, analytical testing and vapor extraction pilot testing and full-scale design were modeled within AutoCAD format using 3-D Studio application software. The results depicted the precise locations of soil layers and the contaminants, and were understandable to the technical staff that used these programs routinely. Additional work was necessary if these were to be made conveniently understandable to a lay group.

The AutoCAD files were transferred to a Macintosh computer for rendering and enhancing the visual appeal. Programs employed within the Mac platform included Canvas, Freehand and PhotoShop. The resulting illustrations retained the technical precision desired by the technical and legal staffs, and gained the easy familiarity needed for a public hearing. The results of soil borings and sampling are presented in Figure 8-1, which depicts the upper and lower surfaces of a silty sand layer that had trapped gasoline contamination. The analytical testing performed on soil samples revealed an oblong contaminated zone as shown in Figure 8-2. Sampling and testing of wells during pilot testing provided the initial extent of well influence zone depicted in Figure 8-3. The final extent of the vapor extraction well influence is illustrated in Figure 8-4.

Output resulting from groundwater transport modeling programs such as MT-3D is some of the most complex and difficult information to comprehend. Visualization programs such as Voxel are able to take this output and construct stunning displays depicting the three-dimensional character of contaminates in the subsurface. Plume maps, estimates of contaminate transport distance, and the prediction of future concentrations at target distances from the source can all be displayed in an easy to understand format.

8.4 HARDWARE REQUIREMENTS AND PREREQUISITE TRAINING

Until recently, a decent-sized workstation, Sun Microsystems or Silicon Graphics, Inc., was the minimum requirement for running geographic database and visualization graphics programs. However, with the availability of the Pentium™ and Power-P.C. chips, desktop systems are being used for some applications. Custom-made Pentium™ computers can do a reasonable amount of processing, and even if they don't have speeds for impressive moving displays, they are certainly capable of file handling and design of individual frame displays. In a client-server environment, even a 486 computer with 32 MB/RAM can be useful, with the data stored on the server. As a general rule, however, be prepared to spend quite a bit, and to upgrade frequently to keep up with cutting-edge technology.

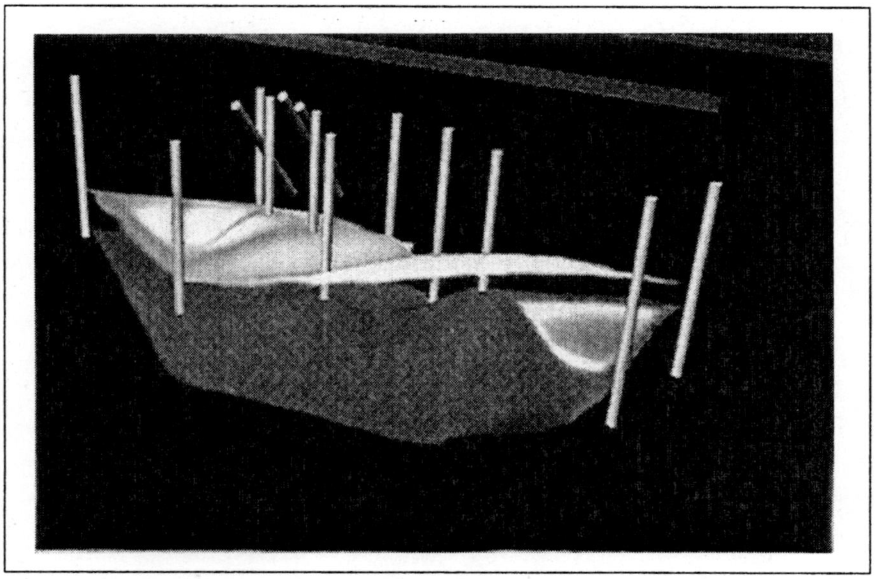

Figure 8-1. Silty Sand Layer - Top and Bottom, with Soil Borings

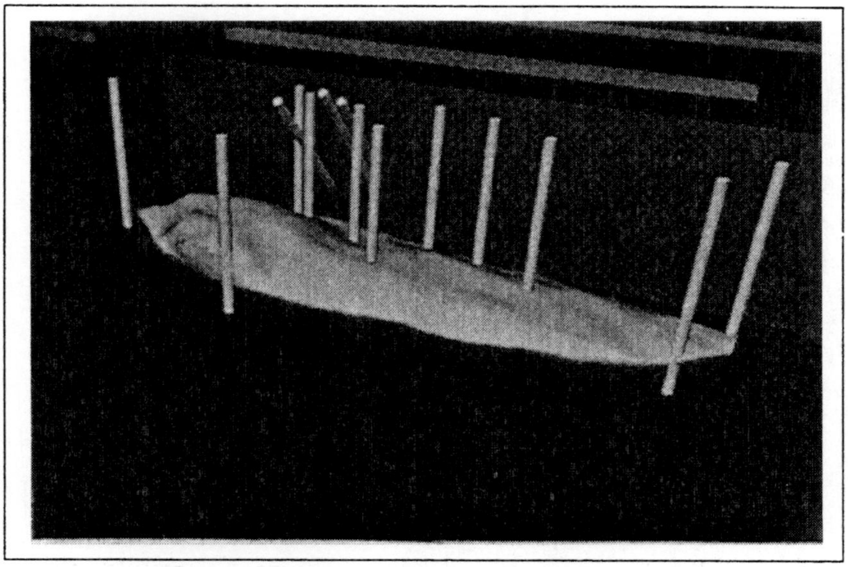

Figure 8-2. Contaminated Zone and Soil Borings

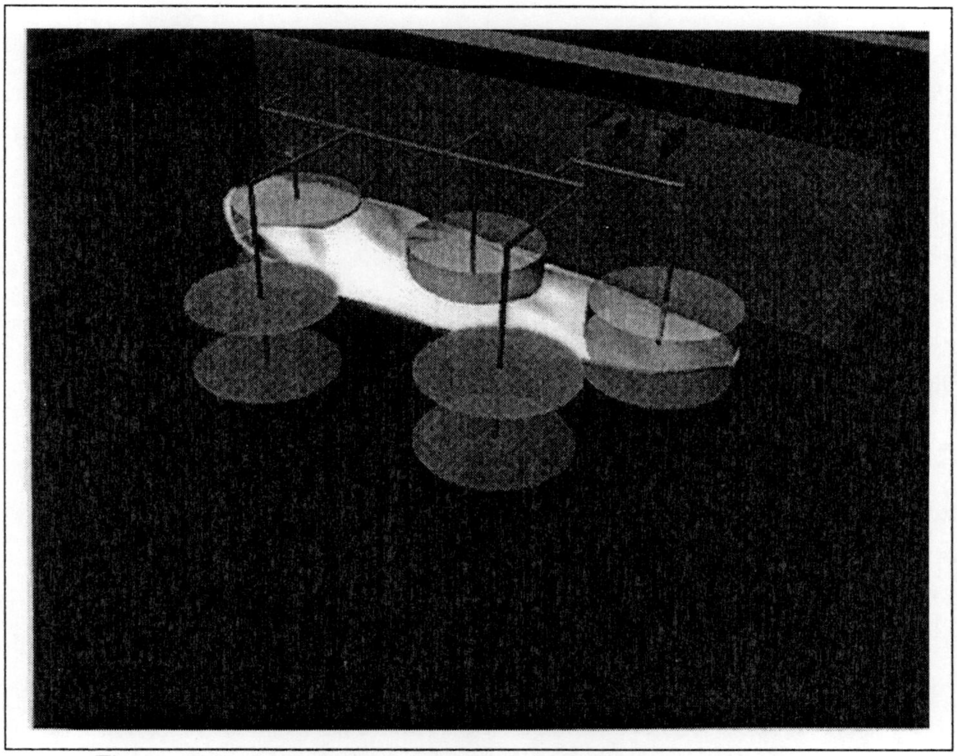

Figure 8-3. Contaminated Zone and Vapor Extraction Wells-
Initial Extent of Influence Zone

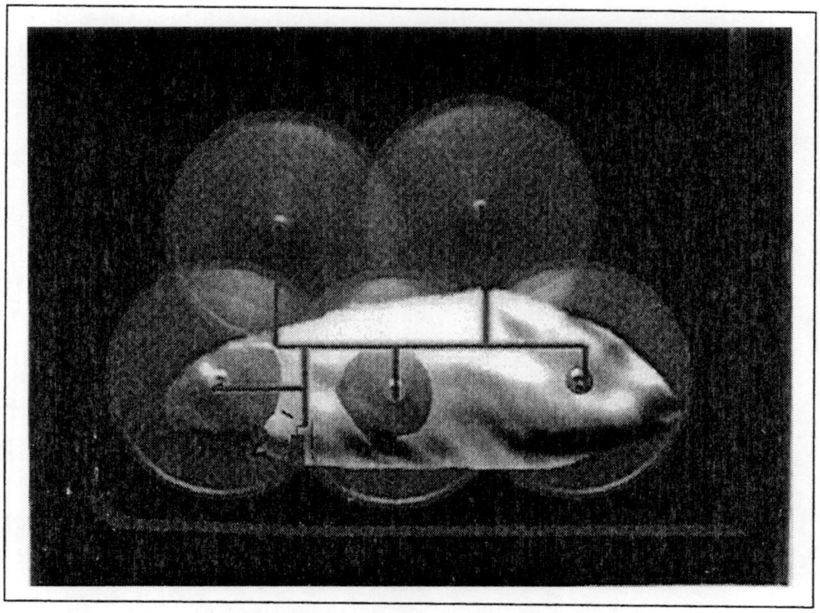

Figure 8-4. Contaminated Zone and Final Extent of Vapor
Extraction Well Influence

The overall combination of program complexity, hardware features and frequent product upgrades becomes quite demanding for the system operator. A proficient operator will require some knowledge of programming languages, an understanding of geology and familiarity with environmental/hazardous waste issues. As a minimum, the operator needs familiarity with the customizing options that each application offers. Also, this individual will need to be full-time or nearly full-time dedicated to the operation and upkeep of the visualization graphics tasks to be able to keep up with upgrades and do maintenance chores such as file backups.

8.5 LIMITATIONS OF DATABASE USE

While at some sites impressive interpretations, displays, and graphics can be created, at other sites the programs and the machines will be unable to make any sense of the data. The database capabilities of computers are impressive, and they allow technical professionals to focus on tasks more important than record-keeping and data reduction. At sites that have confusing geology or visually uninteresting stratigraphy, no amount of graphic data manipulation will add anything to the information as displayed in simple plot plans and tables of analytical data. In such situations, the burden of interpretation remains with experienced individuals who understand structural geology and flow through porous media. Even when the interpretative graphic displays are generated by a computer, there is a distinct difference in raw output resulting from the software and after it has been fine-tuned with the advice of someone who knows how contaminants and groundwater behave in the ground.

8.6 PORTFOLIO MANAGEMENT AND COMPLIANCE TRACKING

Quite distinct from the single-site data management problems of large amounts of investigative and analytical information, another application is the management of the environmental data for property portfolios, including both investigative and compliance data. Application of GIS programs for this application is generally useful for industrial owners, property managers and financial institutions that own large portfolios of real estate for extended periods. The data includes studies that would have been performed at the time of initial purchase, along with any periodic updates that may be done for periodic assessments of environmental liabilities. Individual properties may be "bundled" in various combinations for internal environmental compliance audit, possible sale, or asset review for environmental liability set-aside. A properly maintained database of environmental information on a property portfolio, keyed into the geographic locations of properties, forms a very powerful tool in the hands of decision makers.

A number of off-the-shelf software packages are available, and custom-designed programs can be fashioned from standard components.

CHAPTER 9

HEALTH RISK-BASED ASSESSMENTS

The potential of toxic chemicals to adversely affect human health and the environment is a principal concern at contaminated sites. This is important both during initial site investigation and during development of clean up criteria. Risk-based action level approach for making remediation decisions can be applied to many media (soil, groundwater, air, surface water) and land use considerations (residential, industrial and recreational). This chapter presents an elementary treatment of the subject.

The assumptions, models and methods presented here are based on *the Risk Assessment Guidance for Superfund (RAGS): Vol. I, Human Health Evaluation Manual and Vol. II, Environmental Evaluation Manual (EPA, Office of Emergency and Remedial Response)*. The pivotal question that needs to be answered is: Is the contamination significant from a risk perspective?

9.1 FATE AND TRANSPORT OF CONTAMINANTS

Sources of Contaminants

Contaminants are released to the soil and groundwater from both point (underground tanks) and areal sources (landfills, sumps). Release may be accidental or intentional.

Nature of Contaminants

For purposes of fate and transport analyses, the contaminants can be divided into four main categories.

1. *Miscible Organic Compounds*. These include water-soluble components of petroleum products. Extremely small concentrations of some of these components can affect the odor

and taste of drinking water supplies (for example, phenols or creosols). This property makes their fate analysis very important.

2. *Immiscible Fluids.* (NAPL) The immiscible organic liquids or nonaqueous phase liquids, include several halogenated aliphatics, and tend to lag behind aqueous transport. This property makes it difficult to completely remove hydrocarbons from aquifers.

3. *Miscible Inorganic Compounds.* A number of inorganic compounds, such as nitrates, are found in aqueous media.

4. *Metals.* Metal contamination is commonly found in soil and less frequently in groundwater.

Environmental Factors

A number of different chemical, physical, and biological properties of compounds in conjunction with the environmental nature of the media affect the contaminant transport processes. Some of these are:

Physical: advection, dispersion, volatilization, sorption, and ion exchange.
Chemical: ionization, hydrolysis and oxidation/reduction.
Biological: biomass accumulation and biodegradation.

The physical and chemical factors are well understood. Biomass accumulation is the process by which organisms accumulate and concentrate pollutants. Biodegradation refers to the process of transformation of a chemical by biological agents, usually microorganisms.

Mathematical Modeling

A number of models are available to assist decision-makers in assessing the importance of pollutant pathways, but require considerable judgement. Most models deal with point sources and can be divided into three categories:

1. Vadose Zone Models
2. Groundwater Models
3. Geochemical Models

Either simulation (with empirical equations) or analytical approach (equations describing physics or chemistry of a system) is used to set up a model and a deterministic, statistical or numerical scheme is used for solving the problem.

9.2 DATA EVALUATION AND EXPOSURE ASSESSMENT

Data Evaluation

The first item to be evaluated is whether appropriate type and amount of data required for risk assessment have been gathered. If health-risk based evaluation is applied early enough, it can provide cost effectiveness in data collection.

The main steps in data evaluation are:

1. Select chemicals of concerns in soil and groundwater. This can be based on statistical exceedence from local or regional background concentrations.

2. Perform preliminary identification of potential exposure pathways.

3. Develop a data set for use in risk calculations.

At large complex hazardous waste sites the above steps may be performed after a preliminary investigation and then used to guide a more comprehensive site characterization. The above evaluation can guide the spatial distribution of sampling locations during site investigations with respect to potential pathways.

The spatial distribution may be modeled by geostatistical methods, which provide estimates of representative concentration within target areas.

Exposure Assessment

The next step is to estimate how much, how often, for how long, and at what rate the chemicals can travel from a site to the humans working or living at or near the site. This assessment includes the following steps:

4. Identification and characterization of exposed populations.

5. Evaluation of exposure pathways from sites to exposed population.

6. Estimation of chemical concentrations at points of exposure using environmental fate and transport modeling.

7. Estimation of intake rates in humans from inhalation, ingestion, or dermal exposure to chemicals in air, soil, dust, or water.

EPA's risk assessment guidance requires that risks from a site be assessed in the absence of any remediation, i.e., baseline conditions under both current and plausible future land use scenarios. Human exposure is expressed in terms of intake, defined as the amount of a chemical taken into the body per unit body weight per unit time. Intake rates are calculated separately for each medium (air, groundwater, and soil), and then summed for oral and inhalation routes. If considered significant, dermal exposure is summed with oral exposure. All intakes are expressed in units of milligrams of chemical per kilogram of body weight per day (mg/kg/day).

9.3 TOXICITY ASSESSMENT

This is the process of characterizing the relationship between the dose or intake of a substance and the incidence of an adverse effect in the exposed populations. Toxicity assessments evaluate results from studies with laboratory animals or from human epidemiological studies. At high levels of exposure adverse effects are known, and these evaluations extrapolate to low levels of exposure where effects can only be predicted based on statistical probabilities. Toxic effects are divided into two classes of chemicals: potential carcinogens and non-carcinogens.

The numerical indicators of toxicity are obtained through EPA-developed reference doses (RfDs) or cancer slope factors (SFs). RfD is based on the assumption that thresholds exist for certain non-cancerous toxic effects such as kidney or liver damage, but may not exist for cancer. The RfD is an estimate of a daily exposure that is likely to be without an appreciable risk during a lifetime of exposure.

Cancer is evaluated in two ways: lifetime studies with laboratory animals and human health (epidemiological) studies for exposure to carcinogens. Unless evidence to the contrary exists, it is assumed that any level of a carcinogen has a finite risk of inducing cancer (i.e., zero risk is only associated with zero exposure). Since risks at low levels of exposure cannot be quantified directly from either laboratory animal or human health studies, mathematical models are used to refine the epidemiology results and to extrapolate from high to low doses.

The animal data used are from the most sensitive species studied, based on the assumption that humans are at least as sensitive as the most sensitive animal species. Risk assessments made in this way are health-protective, i.e., the true risk is not likely to be higher than the estimate.

Slope factors for cancer risk define risk due to continuous lifetime exposure to one unit of carcinogen, in units of risk per mg/kg/day. Individual cancer risk is calculated as the product of exposure to a chemical, in mg/kg/day, and the SF for that chemical, in 1/(mg/kg/day). Cancer risks from exposure to multiple carcinogens and multiple pathways are assumed to be additive.

9.4 RISK CHARACTERIZATION

The characterization involves combining the results of the toxicity and exposure assessments. For carcinogens, potential cancer risk is adjusted for lifetime exposure.

Risk = Intake × SF

where, Intake= Chemical intake (mg/kg/day)

SF= Slope factor 1/(mg/kg/day)

To obtain an estimate of total risk from all carcinogens, cancer risks are summed across all exposure pathways.

For noncarcinogens:

Hazard Index = Intake/RfD

where, Intake = Chemical intake, (mg/kg/day)

RfD = Reference dose, (mg/kg/day)

Hazard indices are also summed for those chemicals known to produce similar adverse effects in the same target organ. Allowable exposure criteria are typically based on one excess cancer occurrence in one million exposed individuals.

9.5 ENVIRONMENTAL EVALUATION

This evaluation is generally limited to an identification of potentially sensitive species through a search of Natural Diversity DataBase and analysis of potential exposure pathways to those receptors.

9.6 HEALTH AND SAFETY PLANNING AND MONITORING

Health and Safety of Community and Workers

The purpose of preparing and implementing a health and safety plan is to protect the surrounding community, the site workers and property tenants. Health and safety plans are prepared by certified industrial hygienists, and require some basic understanding of science for implementation. If remediation involving excavation and heavy equipment operation is planned, awareness of occupational safety at construction sites is also needed.

The main danger at hazardous waste sites is from chemical exposure. There is also a secondary danger from potential fire and explosion. In addition to the presence of volatile and toxic chemicals, another hazard to be monitored is oxygen deficiency. This is especially a concern for entry into confined spaces such as basements and large tanks. The low elevation spaces accumulate heavy gases such as methane or carbon dioxide that may crowd out all normal air and oxygen.

At a hazardous waste site the chemicals are often present in an uncontrolled condition. This may include corroded drums, leaking tanks, half-full sumps and abandoned disposal pits. This contrasts with an ongoing factory operation where the same chemicals may be present, but in well-controlled containers and piping. Another factor is the large variety and number of chemical substances. Some of these chemicals may become more dangerous when mixed with or in close proximity to each other. The identity of substances may also not be known, as old labeling may be faded, or chemicals may have been placed in containers intended for some other materials.

Health and Safety Plan

A health & safety plan must describe the anticipated activities in writing and fully address all of the issues. If the proposed activity is site investigation, precautions may need to be taken to protect individuals who will be entering the site, as the nature of chemicals present at the site would not be known. If the plan is to perform remediation, the site will undergo greater disturbance, and a fair amount of information may be available about the chemicals anticipated, thus the plan may be able to specify protection levels that are fully appropriate for the site. To remain current, a health and safety plan should be updated periodically as conditions change or as some of activities at a site are completed.

Essential elements of a health & safety plan include a review of available information, definition of work objectives, methods of sampling the conditions, personnel requirements including the level of training required for the workers. The plan must also include a description of the risks involved, such as from inhalation or skin exposure. The prescribed protection level should be defined, and any required environmental sampling to sample the surroundings should be included. A site map showing the main features, buildings, and known hazards such as waste pits is helpful, in addition to a vicinity map showing routes to the nearest hospital.

Plans will often place some requirements for air monitoring as a measure of the hazardous volatile compounds and dust at the site. In addition to the air monitoring near an open excavation face, or at the site boundary, additional monitoring may be required in the form of personnel monitoring, where an individual's breathing zone is sampled using a handheld or shoulder mounted air sampler. In addition to protection levels for site workers, the plan will define mitigative actions to reduce or control exposure. Decontamination procedures for the site workers will also be included; both for routine daily clean up at the end of each workday, and in case of an emergency or accident.

Chemical Exposure

There are a number of ways in which toxics can enter the body. Inhalation is the most immediate and continuous manner in which chemicals present in the environment affect us. Therefore, respiratory protection is the first and foremost important item at a site where volatile chemicals are present. Ingestion is another route, either indirectly, by eating with unwashed hands, or directly by eating contaminated foods. Drinking contaminated water would also fall in this category. For some very mobile contaminants, dermal exposure is also important. Some of the chemicals can be absorbed through the skin, and one can actually taste them seconds after they touch the skin. A less common way in which toxic substances can enter the body is though puncture wounds in which the skin has been cut. In instances when someone is injured at a hazardous waste site, exposure to chemicals though the open wound is an added danger.

The exposure to chemicals is sometimes acute, causing immediate injury to the receptor. Examples of this type of exposure include tear gas and hydrogen sulfide. At other times the exposure is chronic, and no effects are visible immediately. This is often the case with most carcinogens, which may have a long latency period, and the effects may become evident years later.

The symptoms in the case of many acute exposures are obvious, with burning sensation, coughing, nausea or skin rashes that appear immediately. In other cases, as with most chronic exposures, the exposure is without warning. Effects of asbestos or cigarette smoke may not appear for decades after the exposure begins.

Protection Levels

Personal Protection Equipment for use at hazardous waste sites is prescribed by referring to four different letter designations ranging from Level D—the lowest protection level—to Level A, the highest protection level. Level D is similar to ordinary construction clothing, and includes hard hat, steel-toed boots, and outerwear that may include coverall, goggles and gloves, as needed. Protective outerwear is plain common sense in most situations. The purpose is protection from personal injury due to falling objects, sharp tools, etc., and to prevent any dust or chemicals from adhering to clothing worn at home where children may be exposed or where some of it may be ingested. The outerwear should not be taken home, and should be stored at the site. At some sites disposable coveralls are worn, and discarded on a periodic basis.

Level C, the next higher level of protection includes all of the items prescribed during Level D, plus the use of respiratory protection. Work at most remediation sites is performed under Level D protection, but all workers in close proximity to the contaminated area keep their full-face respirators at hand. These respirators are fit-tested in advance. The workers have to be prepared to upgrade their protection level to Level C by donning their full-face respirators with the appropriate cartridge to filter out the contaminant. The decision to upgrade the protection level is based on air monitoring in the breathing zone of the workers closest to an open excavation face. At sites where liquid contamination is present, chemical-resistant coveralls, gloves and boots are also required.

The next higher protection level upgrade to Level B involves the use of SCBA (Self-Contained Breathing Apparatus) or Supplied Air respirator. This protection level is for use at sites where the concentration of airborne volatile contaminants is so high that cartridge type respirator would be of limited use. The cartridge may quickly become saturated, causing a "breakthrough"—a condition in which the volatile compound saturates the absorbent capacity of the cartridge, leading to an initial whiff of the contaminants' aromatic odor. In case of such a "breakthrough," it is important to immediately leave the hot zone and replace the cartridge with a fresh one. Level B is also prescribed at sites where the nature and concentration of volatiles cannot be anticipated, and an appropriate cartridge cannot be selected, requiring the use of air from outside the hot zone.

The final and highest protection level is Level A, requiring a fully encapsulating suit that will protect not only the site worker, but also the hoses and controls on the SCBA. This level of protection is needed at sites where there is danger of splash from chemicals that may dissolve the hoses or may corrode the control valves. Another addition at this protection level is two-way radio communication. This is because the suit is so confining that it is difficult for the worker to give clear signals or have any meaningful communication with other workers or with personnel outside the hot zone. A summary of the four protection levels is given below.

Level	**Equipment**
A	SCBA or supplied-air respirator
	Fully encapsulating suite, gloves, boots
	Two-way radio communications
	Hard hat
B	SCBA or supplied-air respirator
	Chemical-resistant clothing, gloves, boots
	Hard hat
C	Full-face, canister respirator
	Chemical-resistant clothing, gloves, boots
	Hard hat
D	Coveralls
	Safety glasses
	Hard hat

Hazard Assessment

There are a number of criteria available for deciding when protection against breathable contaminants is necessary, and how much protection should be provided. Some of these criteria are health advisories recommended by NIOSH (National Institute of Safety and Health), while others have the force of law and are issued by OSHA (Office of Safety and Health Administration). Both agencies have adapted or defined a number of different criteria for use by health and safety professionals.

Threshold Limit Value (TLV) by NIOSH

There are three different criteria available:

- TWA (Time-Weighted Average)
- STEL (short-term Exposure Limit)
- C (Ceiling)

Permissible Exposure Limit (PEL) By OSHA

Two criteria are used:

- 8-hour time-weighted average, or
- Ceiling concentration

IDLH Concentrations

Both agencies accept a maximum concentration without protection, with a 30-minute allowance for escaping the site, termed IDLH – or Immediate Danger to Life and Health. This represents an atmosphere that poses immediate hazard to life. If this level were detected at a site, anyone without respiratory protection should leave the affected area immediately.

Decontamination

The purpose of decontamination is to remove and neutralize contaminants that may be adhering to site workers or to the equipment used at the site. This step is necessary whenever an individual or a piece of equipment leaves the hot zone. A decontamination station (or decon station) is set up near the site exit in an access corridor, and includes facilities for personnel and vehicles decontamination.

Items needed at a personnel decontamination station include eyewash, shower, faucet, wash basin, paper towels, wipe cloths and a disposal drum. Any permeable clothing that has been soaked with chemicals will require disposal. Any gross contamination on hands, boots, etc. will require washing, rinsing and wiping. The worker decon station includes lockers where workers may leave their street clothes and shoes during the work hours. The lockers may also be used for hand tools and monitoring equipment that needs to be stored at the end of the work day.

The decontamination station for vehicles and heavy equipment will require long-handled brushes for the vehicle exterior, pumps, pressurized sprays and drains for wash and rinse solutions. It is best to try and remove the contaminated soil without the use of water hoses, as washing may create a lot of contaminated rinsate that requires special handling and disposal.

Site Emergencies

Work at hazardous waste sites should always be considered dangerous, but at times true emergencies will arise, such as injuries or equipment failure. All emergency situations require decisive action on the part of site personnel. Contingency planning for emergencies should include a definition of personnel roles that are well known and fully understood at the site. Training is essential, both for emergency evacuation drills and in administering first aid or CPR. Clear lines of communication can save valuable time when an actual emergency arises. Site evacuation route should be clearly communicated to the workers in advance. As conditions change, the emergency plan should be periodically reviewed and revised as necessary.

Site Work Zones

During site work, it is not necessary to consider the whole site as a hazardous area if the contaminated soil is limited to a portion of the site. It is sufficient to define a hotline around the contaminated work area, and with certain controls, the rest of the site can be free of health and safety requirements. Contaminated area controls are illustrated on Figure 9-1. It would be necessary for anyone entering the overall site to undergo an orientation about the hot zone. However, only those workers entering the limited hot zone would be required to have received health and safety training for hazardous waste operators (HAZWOPER training). The boundary of the hot zone may be as rigid and well defined as a wooden fence, or as flexible as a row of barricades or cones with some caution tape. As conditions change, the zone could be expanded or shrunk down as-needed. The key to safety is regular communication with all site workers about the hot zone boundaries and restrictions.

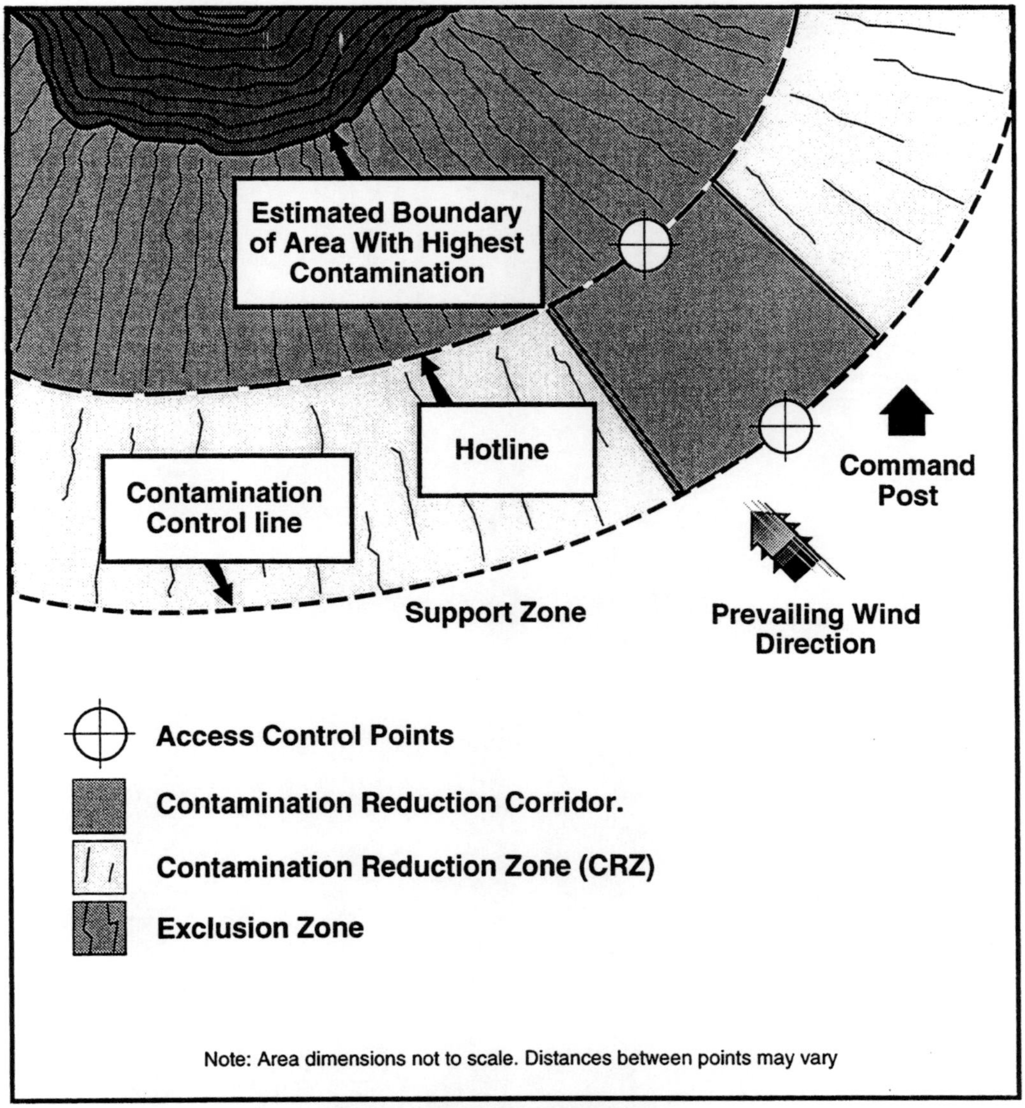

Figure 9-1. Site Work Zones

CHAPTER 10

REGULATORY FRAMEWORK

10.1 EVOLUTION OF ENVIRONMENTAL LAWS

Laws regarding workplace safety and protection from chemicals evolved as the industrial age progressed and exposure to chemicals became commonplace. Most of the worker safety laws were intended to protect factory workers involved in manufacturing. Any exposure of the community, or any harm to the environment was not considered significant. Rachel Carson's 1962 book "Silent Spring" is generally recognized as the beginning of the modern environmental movement in the United States, leading to the passage of Superfund Laws and the establishment of the Environmental Protection Agency. Some of the well-intentioned and progressive environmental laws became overburdened with rigid, deterministic action levels. The single-minded pursuit for assigning liability rather than performing site cleanups has recently begun to change with the use of risk based criteria. This chapter summarizes the laws that form the basis of hazardous waste regulations.

10.2 THE BASIC LAW–RCRA

The Resource Conservation and Recovery Act, or RCRA, was adopted in 1976, and provides the basic framework for federal regulation of hazardous waste. This Act controls the generation, transportation, treatment, storage, and disposal of hazardous waste in a complete manner, termed "cradle to grave." The Environmental Protection Agency has been given the responsibility for administering this law, with major components of the law delegated to the states.

The state programs generally follow the federal program, and the state laws and regulations were designed to complement the federal framework. To be approved, the state program must impose requirements at least as stringent as those imposed under RCRA. Facilities in an approved state do not have to comply with two sets of regulations.

Hazardous Waste

Under RCRA, hazardous waste is defined in one of two ways:

1. It is "listed" as a specific hazardous waste, with a method for de-listing of specific wastes.
2. It has hazardous characteristics, specifically—ignitability, corrosivity, reactivity or toxicity.

Certain classes of wastes are excluded under this definition, including:

- Household hazardous wastes and domestic sewage
- Certain materials used in mining industry or paper pulp industry
- Iron and steel industry coke or coal tar, fly ash and bottom ash
- Drilling fluids and other oilfield wastes
- Samples of soil, air or water that are collected for testing

The states sometimes may regulate more wastes as hazardous wastes than the federal government. Hence, when checking the status of a site, both federal and state regulations need to be checked and compared. Regulatory compliance for generators and transporters includes manifesting and record keeping, maintaining the facilities, groundwater protection, preparedness planning, closure requirements and financial responsibility. Shipping documents called the Uniform Hazardous Waste Manifests are necessary for transporting any wastes. The basic intent of the system is to prevent illegal dumping or other unrecorded disposal of hazardous wastes by creating a paperwork trail.

Generator

A *generator* is any person whose act produces a hazardous waste. A generator must obtain an identification number from EPA before transporting or disposing of a hazardous waste, and when shipping offsite, generators may only use transporters and storage facilities that have EPA ID numbers. Second, the generator must prepare a manifest, and certify that it has selected the safest transporting method for the waste. The generators must also comply with certain pre-transport requirements such as packaging and placarding. The generator must maintain his copy of the manifests for three years, and inform the EPA or the appropriate state agency about the transport. Small *quantity generators*—those producing less than 100 kilograms of waste per month—are exempt from the reporting requirement, and have several other exemptions. If a generator stores hazardous wastes onsite for over 90 days, he must secure a permit as a Treatment, Storage, and Disposal (TSD) facility.

Transporter

A *transporter* is one who moves hazardous wastes from one facility to another. RCRA also imposes certain obligations on Transporters of hazardous waste. They must:

- Obtain an EPA ID number
- Comply with the manifest system, and
- Take action in response to any hazardous waste discharges that occur during transport.

State laws generally add additional requirements for licensing and vehicle registration, which may vary from state to state.

TSD Facilities

Transfer, storage, or disposal facilities such as landfills or recyclers, are subject to extensive regulations under RCRA, including the requirements to obtain an EPA ID number. These facilities are the last link in the cradle-to-grave regulatory chain. When waste is received by a TSD facility, an agent must sign the manifest to acknowledge its receipt and send a copy to the generator. The facility must try to reconcile any discrepancies, and notify an agency of any significant discrepancy. Each TSD owner must obtain a detailed chemical and physical analysis of any wastes to be treated, stored or disposed of.

EPA regulations set forth a number of specific design and operation standards for TSD facilities. One of the features was the introduction of the "land ban," a policy of eliminating land disposal of untreated hazardous waste by 1990. More correctly, this policy in general limited the land disposal to wastes only after they had received some treatment. Some specific wastes were prohibited from land disposal, including landfill disposal of bulk or non-containerized liquid hazardous wastes and dioxin containing wastes. Certain extensions of the land ban dates were allowed for those who could demonstrate that a good-faith effort had been made to locate treatment and disposal facilities in accordance with the original dates. The laws also required EPA to revise its regulations governing waste treatment, as new technologies became available.

Federal and state regulations require TSD facility owners and operators to conduct monitoring of groundwater under their sites. The facility permits designate where the monitoring should be done. The monitoring programs include well installation, sampling, and analytical testing procedures. If the groundwater is found to be contaminated, the regulations specify a protection standard designed to ensure that these constituents do not exceed certain concentrations.

All TSD facilities are required to take corrective action to remedy past and present releases of hazardous waste within their boundaries. The action required is necessary to protect human health and the environment. This requirement was added to the law to force private parties to accept cleanup responsibilities for past releases. Unfortunately, these requirements discouraged some facility owners from applying for RCRA permits. When RCRA requirements are applied to the remediation of wastes at existing contaminated sites, they can act as a disincentive to more protective remedies and can limit the flexibility of the decision-makers in choosing the most protective remedy. Often, the RCRA requirements provided a strong incentive for leaving wastes in place, or for selecting remedies that minimize regulations under RCRA.

Owners and operators of TSD facilities are required to plan for closure and post-closure care of their facilities. At a minimum, closure should be done in a manner so as to minimize the need for further maintenance, and for controlling, minimizing, or eliminating the escape of wastes. This applies to any leachates, contaminated runoff, or waste decomposition products. The facilities are required to prepare plans to address how each unit at the facility will be closed. The owners must also demonstrate the ability to pay for the eventual cost of closure. Therefore, the owners must prepare cost estimates for eventual closure and assure financial wherewithal by establishing a trust fund or by obtaining surety bonding. Owners and operators of surface impondments and waste piles are also required to provide financial assurance.

10.3 THE SUPERFUND – CERCLA

The Comprehensive Environmental Response, Compensation, and Liability Act of 1980 (CERCLA) created national policy to deal with hazardous wastes. The simplified premise of Superfund laws is that the polluter pays. Included in this Act are procedures for containing and removing hazardous waste releases, and for identifying and cleaning up contaminated sites. CERCLA was amended and strengthened by the Superfund Amendments and Reauthorization Act of 1986 (SARA), by expanding the scope of cleanups and the size of Superfund. The two acts collectively are generally referred to as Superfund.

A substantial fund of several billion dollars was created for cleanups, hence the name Superfund. The fund was financed by a tax on crude oil, petroleum products, and by a broad-based tax on business income. Where no responsible party can be found, the federal money pays for 90% of the cleanup, with the states paying for the other 10%.

Superfund's main purpose has been to provide standards and financial assistance for site cleanups and to impose liability on parties responsible for the contamination. In addition, Superfund is also designed to ensure that victims of hazardous substance releases are compensated for their injuries.

Often the responsible parties cannot be identified, or are unable to fund expensive cleanups. In such cases, Superfund provides governmental funding where necessary for remediation of sites.

Most States have created programs similar to Superfund to help finance the cleanup of sites that are not considered a priority under the federal program. Although some sites may be a threat to public health, only the most contaminated sites are included in the federal program.

One main criticism of the Superfund has been that the law places too much emphasis on assigning liability. Thus, much of the money spent to date has been for legal expenses, investigation, and testing, and not cleanups. Initially, a lot of time and energy was spent debating how clean is clean, as many considered the program's standards to be too rigid and stringent. In fact, the program has generated a lot of paper in the form of legal documents and test reports, and not very many sites that have achieved satisfactory closure.

Petroleum Exclusion

Under Superfund, hazardous substances are defined to include RCRA wastes as well as toxics defined under Clean Water Act, the Clean Air Act and the Toxic Substances Control Act. Superfund also provides very notable exclusions for crude oil and petroleum products as long as they are contained in tanks and pipelines. If the petroleum products leak, spill or are otherwise released into the environment, then they are to be dealt with in accordance with the National Contingency Plan and the Clean Water Act.

Reporting Requirements

The first step under the Superfund process is the reporting of any waste releases and the notification by operators of waste dispel facilities. A release of a hazardous substance becomes reportable when it exceeds the Reportable Quantity established by EPA. The Superfund requirements apply to all releases into the environment, rather than just those into the waterways. The reporting or notification cannot be used in a criminal proceeding against the person reporting the information.

Upon notification, EPA or the state agency begins any necessary emergency response or cleanup actions. The lead agency is authorized to undertake removal or remedial action where there is imminent danger to public health, unless the agency determines that the owners of the facility will properly conduct such action. Those liable for the release are liable for costs associated with the removal of contaminated materials. Severely contaminated sites may be evaluated for listing as a Superfund site on the National Priorities List, also known as NPL or the Superfund List.

Cleanup Procedures

In order to implement the cleanups, the Superfund program includes provisions for site investigation, selection of methods for cleanup, and the levels of eventual cleanups to be achieved. EPA or a state agency directs the cleanup operations. Superfund distinguishes between two types of response actions: removal or remediation. Removal refers to a short-term action of limited scope typically carried out immediately after there is a reported release. Other cleanups are termed remedial actions.

The EPA or the state agency has authority for ordering or performing removal actions. The agency can force the property owner to stop the release or prevent future releases. Various factors have been considered in determining whether there has been imminent and substantial endangerment, including the amount, the hazard, the substance released, and the potential for exposure.

Hazard Ranking System and NPL

For remedial actions, the agencies use a Hazard Ranking System, and include candidate sites on the National Priorities List (NPL), which includes abandoned and uncontrolled sites. The list is updated periodically, and excludes sites already included under RCRA, where the operators are required to prevent and clean up contamination under their hazardous waste permits. The Hazard Ranking System is intended to estimate the potential health risks posed by the site through any pathways including air, groundwater, surface water, direct contact, or fire and explosion. Sites which receive a high ranking, are placed on the National Priorities List, which is updated periodically.

RI/FS Process

Superfund defined a very formalized process for selection of cleanup methods, termed the Remedial Investigation/Feasibility Study process, or simply the RI/FS process. RI includes all of the site assessment and characterization activities under which the nature and extent of site contamination and the general site conditions are evaluated. The FS considers the results of site investigation and testing and recommends cleanup method. The cleanup standards have to be selected on the basis of health risk, as well as applicable or relevant and appropriate requirements found in federal, state or local environmental and health laws.

In the feasibility study process, cleanup options are developed and evaluated to select alternatives. Superfund regulations specify nine factors that must be considered in assessing the alternative methods. These are:

1. Overall protection of human health and the environments
2. Compliance with ARARS (Applicable or Relevant and Appropriate Requirements)
3. Long-term effectiveness and permanence
4. Reduction of toxicity, mobility, or volume through treatment.
5. Short-term effectiveness
6. Ability to implement
7. Cost
8. State acceptance
9. Community acceptance

For several years after the passage of Superfund laws, it was customary to apply structured procedures similar to the RI/FS process even to sites that were not on the Superfund list. A lot of time and money was spent in producing the definitive reports on characterization, and in producing voluminous reports on remedial methodologies that were not very likely to be used at a site. Recently, it has become clear that elements of this process are useful, but that a number of alternatives can be rejected out of hand. How well a site is cleaned up is not at all dependent upon the details written about the alternatives that are to be rejected.

Record of Decision

Finally, EPA or other lead agency approves clean up plans and standards in a formal document called a Record of Decision (ROD), which may take years to develop and negotiate. It is not uncommon that by the time the final touches are placed and a ROD is agreed upon, other treatment methodologies have become available, or the comparative costs of various treatment methods have flip-flopped. In many cases, the ROD has to be rigidly adhered to by all parties.

Potentially Responsible Parties

Superfund included extensive provisions for the identification of responsible parties. Before any liability can be actually proved, EPA and the states can name Potentially Responsible Parties (PRP's), who may be required to finance cleanup activities if they are found to be responsible. A PRP may be any person who currently owns or operates a facility where hazardous substances were released. Past owners and operators are also included, as well as those who arranged for or actually transported, treated, or disposed of the hazardous substances to the site. The Superfund laws only require that there has been an actual release, negligence or other wrongdoing is not required. Parties identified may have to pay cleanup costs, even if they had followed all regulatory requirements at the time of disposal or past release. Many small and large businesses that had the misfortune of having purchased contaminated sites for their facilities, or had operated them in the past, were driven out of business while trying to clean up the properties. Superfund liabilities are so broad that even sates and local governments have been held responsible for cleanups due to their past practices. The law also applies retroactively to parties who operated facilities before the passage of the laws.

Under Superfund, responsible parties are ultimately liable for all costs of a removal or remedial action incurred by the federal or state governments, or any other person, and any damages for injury, destruction or loss of natural resources. Any person found liable for response costs may seek contribution from other liable or potentially liable parties. A court may allocate such costs among responsible parties as it determines to be equitable. All this emphasis on assigning liability has resulted in considerable legal expenditure while the potentially responsible parties defend themselves from EPA, and sue each other.

"Innocent Buyer" Defense

Superfund provided an important defense termed the "Innocent Landowner" defense or the

"Innocent Buyer" for property owners who acquire land and subsequently discover that hazardous substances were disposed of on the property without their knowledge. The property owner must establish that he had no reason to know of the presence of hazardous substances. As awareness of the likelihood of site contamination has spread, the viability of this defense in the absence of any pre-purchase investigation has become limited. Some states have passed laws that allow the government to impose a "super lien" on the responsible party's assets, with priority over all other claims.

It has become standard practice for purchasers to perform site assessments (as discussed in Chapter 3) prior to real property transfers in order to qualify as "Innocent Buyers." If contaminants are confirmed or discovered, the property transfer agreement and the price may be significantly affected, or the deal may fall through. A buyer may still be willing to purchase the site, provided that remediation is performed as a condition of the sale.

10.4 UNDERGROUND STORAGE TANKS (UST)

In 1984 Congress passed a set of laws dealing with the underground tanks, and separating the tanks from the RCRA laws. This simplification has led to much expedited cleanup of underground fuel storage tanks. At many street corners one can see gas station tanks that are being excavated and replaced, and the sites cleaned up, while only a fraction of the facilities under the Superfund or RCRA have ever been able to achieve closure. Many states drafted their own laws for underground tanks, and some, such as California, even preceded the federal programs.

The Federal law defines UST as a tank that is used to store regulated substances, and 10% or more of the tank's volume is below the surface of the ground. Not included in the UST's are non-commercial farm or residential tanks containing fuel or heating oil and with a capacity of 1,100 gallons or less. Also not included is a storage tank located in a basement, if the tank is entirely upon or above the surface of the floor. EPA's regulations also exclude UST systems storing hazardous wastes regulated under RCRA. Thus, fuel tanks at a site may be handled according to simpler UST laws, while a waste oil tank may have to be removed using the much more complicated RCRA laws for closure.

The federal and state laws require all owners of UST's to register, providing information on the age, size, location and uses of the tank. In many communities, the registration requirements have been delegated to the local city or county fire departments, health agencies, or public works departments. All newly installed USTs must also be registered. Tanks for wastes must be double lined, although simple fuels may still be stored in single wall tanks. After installation, the tank owners must comply with general operation requirements for ongoing spill and overfill control procedures, corrosion protection and compatibility of the UST with the product.

The Federal UST Law provides for reporting, investigation and confirmation of any releases. EPA's regulations apply these requirements to spills and overfills above certain threshold reporting requirements. State laws contain their own reporting requirements, including telephoning the relevant agency and subsequent written reports. Most states require installer permits and certification for anyone installing modifying or removing a UST.

For permanent closure, a UST must first be emptied of all liquids, rinsed, and then either removed from the ground or filled with an inert solid material. Federal law also includes provisions for a fund and a mechanism for Leaking Underground Storage Tanks, termed a LUST fund.

10.5 CLEAN WATER REGULATIONS

The original Federal Water Pollution Control Act was passed by Congress in 1948, and amended a number of times, with the most significant being the Clean Water Act (CWA) of 1977. States have developed their own laws and programs for water pollution control. The Clean Water Act applies to all navigable waters of the United States. The federal Safe Drinking Water Act addresses aspects of groundwater protection. The state water pollution control laws generally define waters of the state to include groundwater. For instance, California's Porter-Cologne Act applies to any water, surface or underground, including saline waters. New Jersey's Water Pollution Control Act applies to the ocean and its estuaries, all springs, streams and bodies of surface or groundwater.

To regulate point source discharges, the Clean Water Act outlines a nationwide permit system, known as National Pollutant Discharge Elimination System, or NPDES. Point Source is defined very broadly and includes erosion runoff, or the mere placement of pollutants so that precipitation causes a discharge into navigable waters. This broad definition poses severe ramifications for hazardous waste storage and disposal facilities. NPDES requires a person to obtain a permit prior to discharging any pollutant into the waters of the United States from any point source. It may be obtained by applying to the EPA or to the state if it has a certified Clean Water Act program. NPDES permits are effective for terms not exceeding five years. The permit system is intended to establish the level of performance that the discharger must maintain. The permits include effluent limitations, reporting requirements and a schedule for compliance.

Like other dischargers, Publicly Owned Treatment Works (POTWs) are also required to obtain NPDES permits. The treatment works in turn impose pretreatment standards on industrial users in order to ensure compliance with their own NPDES permits by preventing the introduction of pollutants that might interfere with equipment or operations.

A special permit must be obtained from the Army Corps of Engineers prior to the discharge of dredge and fill materials into the navigable waters. This permit is referred to as the "Section 404" permit.

10.6 ASBESTOS REGULATIONS

Asbestos is the common name for a group of naturally occurring minerals that have fibrous, needle-like crystals. Two of the commonly occurring members of the group are amosite and chrysotile. Alchemists and others have used asbestos-containing materials (ACMs) throughout history as fire protection materials in various ceremonies and experiments. Beginning in the 1930's the availability and use of asbestos became very widespread, and some percentage of the material was being included in many applications, including spray-on insulation, insulation around pipes, ceiling and floor tiles and in cloth and rope. EPA defines ACM as any material that contains at least 1% asbestos.

The main hazard of asbestos is that it can be inhaled easily. Asbestos particles get lodged in lungs, where they irritate the lung tissue, eventually developing into lung cancer. Smokers with asbestos in their lungs are particularly susceptible to lung cancer. However, the adverse effects of asbestos have a long latency period, sometimes taking up to twenty to thirty years to become fully evident. In order to be inhaled, the asbestos has to be in a friable state, so that it can be pulverized by hand pressure.

By the mid 1970's EPA had banned the use of asbestos in the manufacture of building materials. However, some buildings constructed as late as the mid 1980s used asbestos containing materials that were manufactured ten years earlier. EPA never issued any requirements for removal of ACM's from existing building. However, OSHA established a standard for permissible airborne exposure of 0.1 fibers per cubic centimeter of breathable air. Thus, employers and building owners had to perform abatement of asbestos containing materials in their buildings. OSHA's requirements for the construction industry included proper respiratory protection, protective clothing, hygiene facilities and guidelines on controls for asbestos removal, renovation and demolition operations.

When a building is demolished, or asbestos materials are removed, advance notice must be filed with EPA or the appropriate state agency. ACM's can only be removed with wet removal techniques. No visible emissions of dust are allowed during removal, transportation or disposal of ACM. Removal of asbestos from buildings is still not required, nor are ACM considered a hazardous waste under RCRA or CERCLA. However, many state and local governments have very stringent regulations. California, for instance, has declared asbestos as a hazardous waste.

Children have the longest life expectancy, and as asbestos hazard has a long latency period, the children are at the greatest risk. Children also spend a lot of their time in school, where asbestos is susceptible to disturbance by the occupants. Therefor, EPA has focused on the problem of asbestos is schools. The Asbestos Hazard Emergency Response Act (AHERA) requirements pertain to schools and were enacted in 1986.

10.7 PCB REGULATIONS

Polychlorinated biphenyls (PCBs) are a group of about 200 chemicals that have been used in transformers, coolants and throughout the power industry due to their extremely stable nature and heat resistance. EPA defines PCBs as oils or other materials containing 50 PPM or greater of polychlorinated biphenyls and the Toxic Substances Control Act of 1976 has banned their manufacture. It is estimated that due to their extremely stable nature, the over one billion pounds of PCB that have been manufactured are still present in the environment.

10.8 RADON REGULATIONS

Radon is a heavy, colorless, odorless, radioactive gas that is formed by the decay of radium. It occurs naturally in rocks and soil containing granite or shale and was commercially used in luminescent products. Radon-related products are a lung cancer risk, and typically occur in confined spaces in contact with ground, such as basements and cellars. At present the action level is 4 picocuries per liter of air. Remediation methods include placing a barrier to prevent radon from seeping into the enclosed area, or increasing ventilation.

10.9 INSECTICIDE, FUNGICIDE, AND RODENTICIDE ACT

The Federal Insecticide, Fungicide and Rodenticide Act (FIFRA) of 1972 mandates that all pesticides intended for sale should be registered. These regulations also authorize states to set standards and establish certifications. Some pesticides regulated under FIFRA are also considered toxic pollutants under the Safe Drinking water standards.

10.10 CLEAN AIR ACT

The Clean Air Act created the framework for protecting and improving the nation's air quality. The Act directs EPA to set air quality standards and to set emission limitations. It also has special provisions for hazardous air pollutants. The National Ambient Air Quality Standards have set criteria for the following pollutants: carbon monoxide, lead, nitrogen dioxide, ozone, sulfur dioxide, and particulates. Some states, for instance California, and the northeastern states have adopted more stringent standards.

EPA initially regulated emissions of seven hazardous pollutants under the National Emission Standards for Hazardous Air Pollutants (NESHAPs) program: asbestos, benzene, beryllium, inorganic arsenic, mercury, radionucleides, and vinyl chloride. This list was later expanded to cover additional substances. The 1990 amendments also mandated the phasing-out of substances that contribute to the depletion of the ozone layer, and replacing them with substitute products.

CHAPTER 11

CONFIRMATION SAMPLING AND SITE CLOSURE

11 .1 GENERAL

Confirmation sampling refers to the sampling and testing performed after remediation and cleanup at a site. The purpose of confirmation sampling is to ascertain if the site has been adequately remediated. The techniques and methods employed during confirmation sampling are the same as those employed during site investigations. However, the consequences of inaccurate or nonrepresentative sampling or testing are much more severe. Inadequacies of sampling or testing during a site investigation can be ameliorated simply by extending the investigation and obtaining a few more samples to fill data gaps, or by taking repeat samples. In comparison, errors during confirmation sampling after remediation may lead to erroneous closure of a contaminated site or additional remediation when none is needed.

In view of this import, confirmation sampling programs are undertaken in accordance with formal plans that require regulatory approvals, and the results are subjected to a lot of scrutiny. Often, a confirmation sampling plan is prepared and agreed upon even before remediation is undertaken. Having a predefined and preapproved confirmation sampling plan allows the remediation manager to perform intermediate checks on the remediation process to assess that the methods employed will achieve the final results. The method of remediation may be modified in order to satisfy the final closure criteria.

The principles introduced in the following paragraphs are common to an initial site investigation and a site closure, but all references in the following text are to site closure.

11 .2 SAMPLING PRECISION AND ACCURACY

Definitions

Precision is the closeness of repeated sample values. In the context of laboratory testing, if a test performed on a particular sample is repeated, the closeness of test results indicates the precision. Sampling precision refers to the repeatability of the sample results. In the case of samples of contaminated soil or groundwater obtained from a site undergoing remediation, both the sampling and test methods have to be repeatable in order to achieve the desired precision.

Accuracy refers to the closeness of a sample value to the true value for the site. This statement implies that the true value for the site is something that can be determined. Therefore, when talking about accuracy, we are referring to a standard or a norm that may not be known at the time of sampling. With this premise in mind – that the true value is a determinable quantity–accuracy is a criterion that must be satisfied during confirmation sampling. The concepts of Precision and Accuracy are illustrated in Figure 11-1.

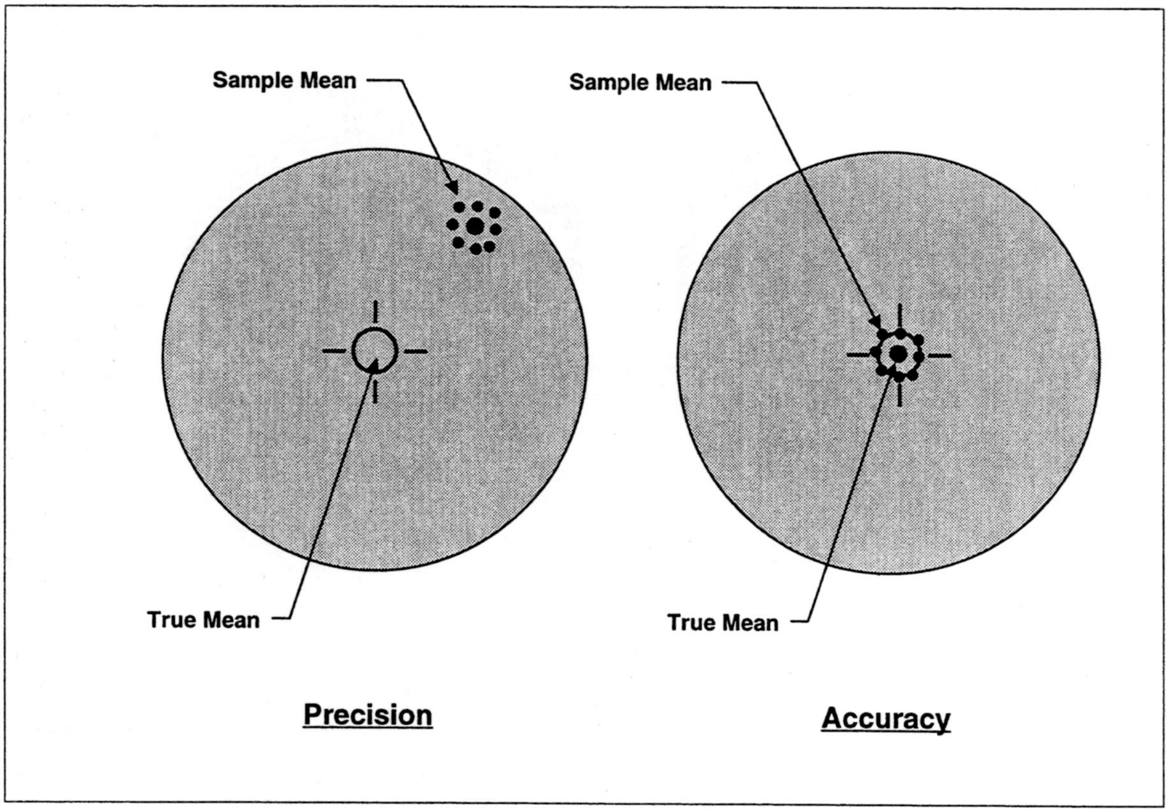

Figure 11-1. Precision and Accuracy in Sampling

Improving Precision and Accuracy

Precision can be improved by increasing the number of samples. This improvement is achieved up to a point and as the number becomes very large, further improvement is minimal. Lack of precision is easy to identify, as the sample results vary over a broad range, showing scatter in data. Precision is also improved by trying to remove variability in operators, laboratories, sampling and testing equipment, sample size and sample handling procedures. If the same person obtains all samples using the same procedure, precision is likely to be greater than if different operators and methods are used. Similarly, performed tests at one laboratory provide consistency and increase precision. Sample size (weight or volume), if kept consistent, will also yield more repeatable results.

Field procedures such as time of the day when a soil sample is obtained, whether it is stored somewhere before being taken to the laboratory or directly transported for testing, are all factors that influence precision. Even after arrival at the laboratory, if the storage methods, holding time and other practices are kept consistent, higher precision is achieved. Maintaining consistency and better practices in the field during sampling and good practices in the laboratory all increase precision. Many of these steps also increase the expense of sampling and testing. In other words, increase in precision may increase the cost of sampling and testing.

While low precision is easy to recognize, the lack of accuracy is not obvious, since the true value for a site is not known at the outset. Achieving higher accuracy is not a matter of increasing sample size or using the same operator for testing the samples. Increasing the number of samples also does not automatically increase the accuracy. Whereas low precision may reduce costs, low accuracy does not lead to any savings. Similarly, achieving higher accuracy does not necessarily cost money, but is a matter of better understanding and planning.

Accuracy is achieved by performing representative sampling of the site. Representative sampling is a matter of understanding the population to be sampled, such as the distribution of contamination at the site, as described in the next section.

11 .3 REPRESENTATIVE SAMPLING

Designing a representative sampling program for a population, or a site, requires some knowledge of the site. If the population has a normal distribution of the parameter to be measured, then a random sampling program is required. In case of a contaminated site, this would mean that the chemical concentrations follow a normal distribution pattern–a few points at the site have extremely high concentrations and a few points are almost clean, while most of the points at the site have contaminant concentrations that fall somewhere in between.

A normal distribution can be represented by a bell curve (Figure 11-2). A sampling plan to obtain random samples will be appropriate for this site.

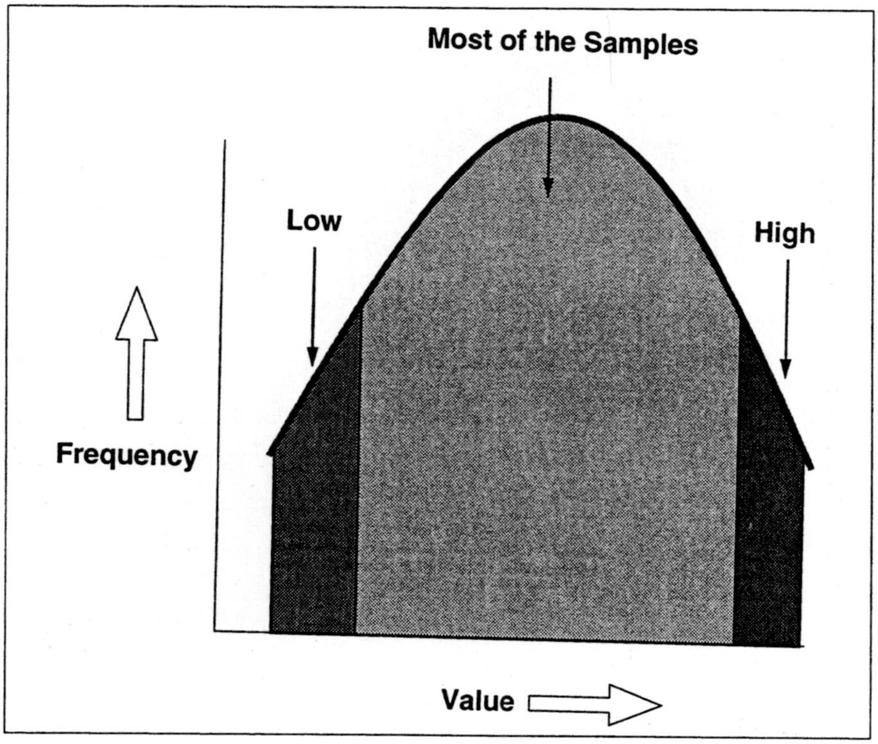

Figure 11-2. Normal Distribution

Hot Spots

If, however, we know that certain hot spots or zones of high concentration exist at the site, then the overall distribution may not follow a normal distribution. It would be best to design a separate sampling plan for these areas, thus effectively removing them from the population to be sampled, and making the remaining of the site into a normal population. The hot shots would first be sampled using a focused (biased or authoritative) sampling program. This kind of sampling is best guided by an individual knowledgeable about the hot spots. These may have been areas that were identified during the site inspection, or areas that became apparent during remediation. Site history may also provide guidance, and may indicate such hot spots as areas of known past disposal, spills, storage of chemicals or operation of machinery. These areas may all be known to, or suspected of, having unusually high concentrations that fall outside normal distribution. An authoritative (nonrandom) sampling program can be carried out as follows:

- An individual well-acquainted with the site and with the contaminants of concern selects "Hot Spots," and the sampling locations.
- Samples are then obtained from each hot spot, and a record is kept of the boundaries of the hot spots or zones of nonrandom contamination Figure 11-3.
- Supplemental remediation may be needed for all hot spots, depending upon the test results; therefore, this method should be used with caution during confirmation sampling.

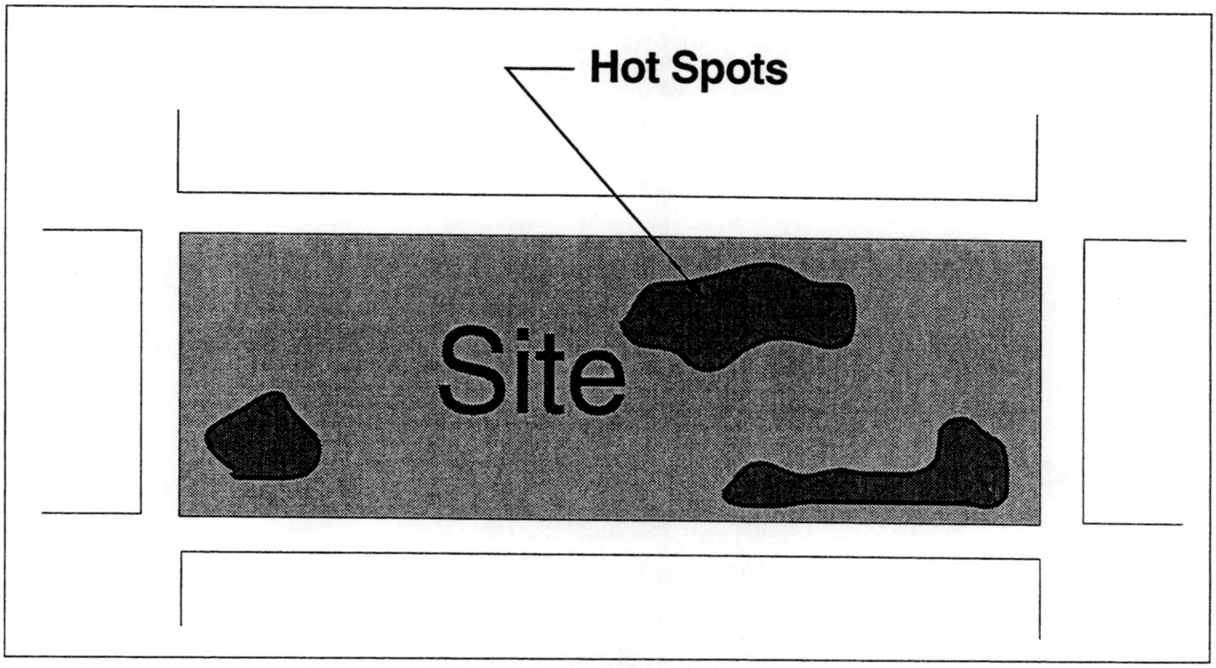

Figure 11-3. Hot Spots on a Site

Random Sampling

The remainder of the site can then be sampled by employing a random sampling program. Random sampling of a site means that every point at the site has an equal chance of being sampled. Random sampling is not haphazard sampling. Sampling performed in accordance with a preconceived plan–such as at all corners, or at all third-points–is also not random sampling.

For selecting random samples, a site is divided into grid blocks which are numbered. Every grid block should have an equal chance of being sampled. A site that is only contaminated at the surface, such as former farmland with pesticides, can be divided into equal-sized squares on an imaginary grid. For subsurface samples, such as for underground fuel tank investigations, the zone of concern may be defined as a layer or stratum where contaminant of concern is found or expected, requiring simple random sampling. If several contaminants at several levels were to be sampled, stratified random sampling would be needed.

After subdividing a site into grid blocks, the number of samples, dependent on the desired confidence level, is selected. Random number generators or a random number table is used to select the squares, or zones, that are to be sampled. This sampling process is illustrated by means of an example in the next section.

11 .4 CONFIDENCE LEVELS AND GEOSTATISTICS: EXAMPLE PROBLEM

When designing a sampling program, there is a need to define the confidence level desired. If a site is suspected of containing an abandoned underground fuel tank, an investigation may need to be able to state with, say 80% confidence, that the underground tank has, or has not leaked. If a high confidence level is desired, the sampling program will need to be more elaborate. Similarly, if the item being sought is smaller–say an abandoned drum rather than a large UST–a greater number of points will need to be sampled.

Geostatistics–the statistics of earth media–can be used to express the understanding of a site in terms of confidence levels. Data already obtained for a site can be examined to see what kind of confidence level has been achieved. One example of geostatistical evaluation is EPA Method SW-846. This method is used in the example given below.

Site Background: Preliminary sampling at an underground storage tank site shows that Total Petroleum Hydrocarbons measured as gasoline are present at or near the regulatory threshold. The hydrocarbons are present within a narrow stratum, which is continuous across the site. The distribution of hydrocarbon concentrations across the site is considered to be random, as the site soils have been disturbed by excavation and backfilling. Test results on four preliminary samples are as follows: 86, 90, 98 and 104 mg/kg. Regulatory Threshold for gasoline contamination is 100 mg/kg.

Problem Statement: Design a confirmation sampling program to achieve an 80% confidence level in our knowledge of hydrocarbons present at the site. Select the number and locations of additional samples to be obtained.

Analysis of Preliminary Samples:

- No. of preliminary samples, n = 4

$$\overline{X} = \frac{\sum_{i=1}^{n} X_i}{n} = \frac{86+90+98+104}{4} = 94.50$$

- Mean of the preliminary test result,
- Degrees of Freedom = number of preliminary samples less one = 4-1= 3

- Variance (Square of Standard Deviation), s^2

$$s^2 = \frac{\sum_{i=1}^{n} X_i^2 - (\sum_{i=1}^{n} X_i)^2 / n}{n-1} = \frac{(86^2+90^2+98^2+104^2) - \frac{(86+90+98+104)^2}{4}}{3} = \frac{35{,}916 - 35{,}721}{3}$$

$= 65.0$

ρ = Difference between Regulatory Threshold and the Mean of Preliminary Samples
= 100 - 94.50 = 5.50

No. of Additional Samples: We will employ Student's "t" method for calculating the number of additional samples required.

- Regulatory Threshold = 100 mg/kg;
- Mean value for preliminary samples = 94.50, and,
- Variance for preliminary samples = 65.00,

The number of Additional Samples required to achieve 80% confidence level, n_1, is:

$$n_1 = \frac{(t^2_{20})(s^2)}{\Delta^2} = \frac{(1.638^2)(65.00)}{(5.50)^2} = 5.77, \text{ say 6 additional samples.}$$

Where $t^2{}_{20}$ is obtained from tabulated values shown in Table 11.1, based on degrees of freedom for the preliminary samples = 3, prepared for Students"t" method.

Selection of Sample Locations: The contaminated zone at the site is within a narrow stratum and contains only one contaminant, gasoline. Therefore, simple random sampling will be done using the following steps to select sample locations.

- ❑ Using a site plan, divide the site into a grid, assigning sequential numbers to all grid blocks: Figure 11-4 shows grid blocks numbered from 0 to 107.

- ❑ Select 6 grid blocks for sampling using random number tables or a random number generator provided in many spread sheets or statistical programs and calculators. For example, Lotus 123 command for random number generator is: @RAND(107). Each time this command is used, the program will provide one random number between 0 and 107.

- To obtain six grid block numbers simultaneously, set up the equation six times on your PC screen. Each time a number is entered, six new grid points between 0 and 107 will be displayed.

TABLE 11-1
TABULATED VALUES OF STUDENT'S "t"
FOR EVALUATING SOLID WASTES

Degrees of Freedom (n-l)[a]	Tabulated "t" Value[b]	Degrees of Freedom (n-l)[a]	Tabulated "t" Value[b]
1	3.078	18	1.330
2	1.886	19	1.328
3	1.638	20	1.325
4	1.533	21	1.323
5	1.476	22	1.321
6	1.440	23	1.319
7	1.415	24	1.318
8	1.397	25	1.316
9	1.383	26	1.315
10	1.372	27	1.314
11	1.363	28	1.313
12	1.356	29	1.31j1
13	1.350	30	1.310
14	1.345	40	1.303
15	1.341	60	1.296
16	1.337	120	1.289
17	1.333	00	1.282

a Degrees of freedom (df) are equal to the number of samples (n) collected from a solid waste less one.

b tabulated "t" values are for a two-tailed confidence interval and a probability of 0.20 (the same values are applicable to a one-tailed confidence interval and a probability of 0.10). If duplicate numbers appear, the process can be repeated. Any of the repetitions is statistically valid. An illustration is shown in Figure 11-6.

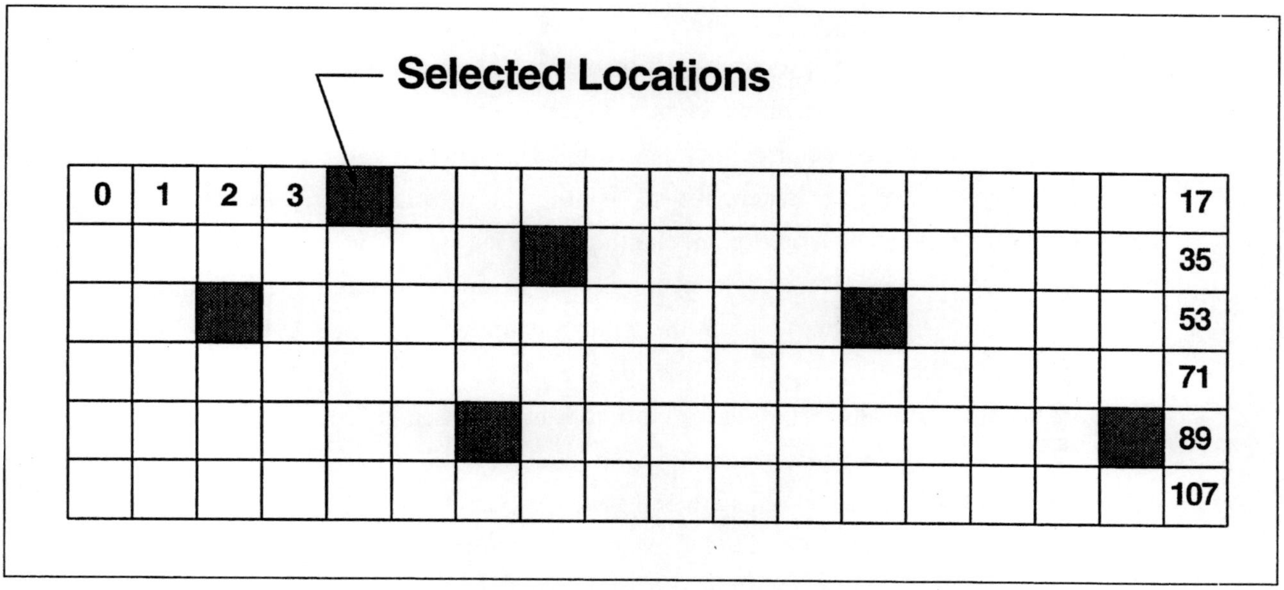

Figure 11-4. Grid for Random Sampling

RUN NO. 1	**48**	Area Size: 180 feet x 60 feet
RUN NO. 2	**88**	Grid Spacing: 10 feet
RUN NO. 3	**4**	Total Grid Sampling Points Possible: 6x18=108
RUN NO. 4	**78**	No. of Samples Desired=6
RUN NO. 5	**25**	Lotus 123; @RAND*107
RUN NO. 6	**38**	

Figure 11-5. Random Sampling Plan

11 .5 REGULATORY THRESHOLDS AND ACCEPTANCE CRITERIA

The objective of remediation at any site is to lower the contaminant concentrations at the site to within the applicable regulatory thresholds or below the criteria recognized by the lead regulatory agency.

These criteria could be derived from health-risk based analyses, or generally accepted background values applicable to the region. Agreed upon values of these thresholds or criteria become critical in designing, starting and, shutting down the remediation process at a site.

The design of a remediation program must incorporate the particular regulatory thresholds to be satisfied. For instance, the performance of a Vapor Extraction System during operation may be evaluated by testing the concentration of vapors at the outlet. This data may also be provided to the regulatory agencies that have oversight responsibility. However, the vapor concentration may not be permissible for use as the criterion for shutting down the system or for declaring that the site has been cleaned up. The basis of acceptance may solely be the concentration of contaminants in soil samples obtained by drilling within the plume boundaries. If concentrations in the vapor measurements are dropping off, one may decide to drill some borings and obtain samples for testing. One does not have to wait until all vapor emissions have ceased before doing the soil testing.

Similarly, if contaminated materials are being removed from the site by excavation, the location, number and type of samples tested in the excavation bottom and walls may be crucial to knowing when to stop digging further. It may be important to know when to pause in the excavation so the required sampling can be fulfilled. The testing of excavated material also becomes important, as some of the material may be below the regulatory threshold, and may not require remediation. This material could even be placed back in the excavation as clean fill. Apparently clean material, if segregated into stockpiles based on appearance, odor, hand-held instrument readings or field test kit results, will require further analytical work for proving that a particular stockpile is clean.

The relationship between regulatory threshold value and the accuracy or precision of confirmation sampling effort is illustrated in the Figure 11-6. A case where the confirmation sampling is accurate but imprecise is shown on Figure 11-6A. The sample results have a wide scatter, and yielded a very broad bell curve. However, since the sampling was accurate, the mean of the widely scattered sample and the true site mean coincided. The imprecision leads to an erroneous conclusion that the site is contaminated, since the regulatory threshold falls below the upper limit of data scatter. Unnecessary remediation will likely continue at this site.

In contrast, Figure 11-6B shows confirmation sampling that is inaccurate but precise, yielding a very narrow bell curve. In this case, the site is erroneously considered clean. The upper limit of the confirmation sampling results falls below the regulatory threshold. Further sampling or checks performed by a regulatory agency could show that inadequate remediation has been performed in Case B. In either case, a better understanding of the regulatory threshold and better sampling could have prevented the false outcome.

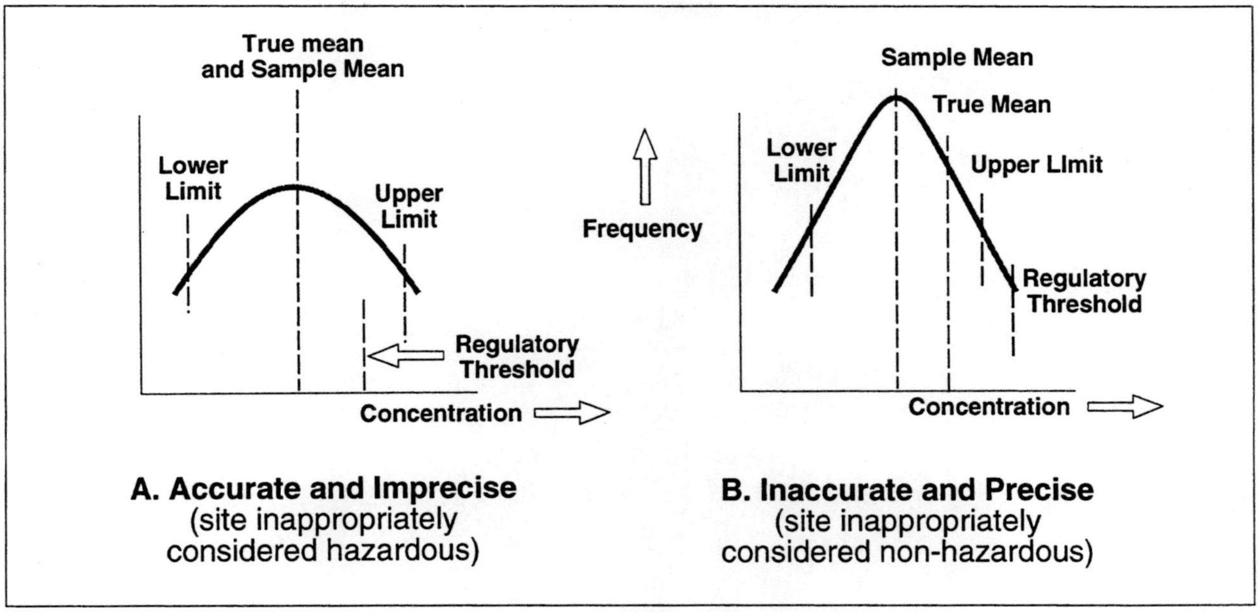

Figure 11-6. Precision, Accuracy and Regulatory Thresholds

11 .6 SITE CLOSURE AND AGENCY SIGN OFF

Site closure does not have to be a complete site cleanup. In fact, as our ability to measure smaller and smaller concentrations increases, it is becoming virtually impossible to declare any site "clean." Fortunately, the regulatory developments are leading us beyond rigid deterministic criteria and towards more flexible criteria based on health risk. A project manager trying to obtain closure at a site should realize the fact that the site is not being cleaned up, but merely remediated. Closure reports, correspondence and other statements should avoid mentioning that the site has been, or will be, cleaned up.

In many instances, there may not be a need to obtain a very clearly defined site closure, for instance if the property is not changing ownership, and its general use—say industrial—is continuing. Also, there may not be a need to obtain a written closure from an agency, but merely a tacit approval of no further action. Such an approval may not even require correspondence issued by an agency but may be based on documentation prepared by the project manager and sent to the agency by certified mail. No agency reply may be received, but if a tacit approval is given, further approval may not be necessary to carry out the development plans by the same industrial owner.

However, if the property is to be sold, and a quick, clear and unambiguous closure written by an agency is necessary, then perhaps a different method of remediation, such as excavation and off-site disposal, should be selected. In short, the final end point—site closure—should be kept in mind when selecting the remediation method and designing a confirmation sampling program.

11.7 APPLICATION TO BROWNFIELDS

Brownfields are urban properties that have a strong market potential for development. Projects proposed for these sites make sense for the community and are accommodated within the current zoning. These sites have just one problem: the properties are labeled "contaminated." The cost of the clean up needs to be reasonable and controlled. Concepts covering health risk and geostatistics introduced in Chapters 9 and 11 can be helpful in lowering the cleanup costs. Rather than blindly following deterministic criteria, creative options can be selected on the basis of principles stated here.

Case History: Brownfield Site

The former US Steel manufacturing facility in south San Francisco is under mitigation for development as commercial and retail lots. The site, which had been in use from the 1930s through the mid 1980s, is comprised of approximately 47 onshore and 120 offshore acres. Site characterization studies identified site-wide soil contamination with hydrocarbons and heavy metals, extensive sediment hydrocarbon and metals contamination in near-shore areas, and shallow soil contamination onshore. Approximately three discontinuous acres of Section 404 wetlands were also identified at the site.

The project environmental consultant, Treadwell & Rollo, Inc. of San Francisco, developed both on- and off-shore characterization and mitigation programs. The on-shore program involved the identification of seventeen soil contamination "hot-spots" and the selection of hydrocarbon and metal mitigation plans. A health and ecological-based risk evaluation was performed to set cleanup standards for soil within 100 feet of San Francisco Bay and across the remainder of the site.

Approximately 10,000 cubic yards of lead-contaminated soil were excavated and treated using stabilization with a 5 to 10% by weight Portland cement mix. The soil was treated on-site to reduce the cost of handling and transport, and subsequently covered with clean fill. Over 350,000 cubic yards of clean fill were imported to the site as cover material for contaminated soil and to bring the site to final grade.

To deal with contaminated sediments, an innovative approach including targeted dredging, sediment capping, sediment disposal, and wetland construction was developed to allow site mitigation and development to proceed. Areas identified as containing contaminants which may be bioavailable will be dredged and capped with clean sediment, with the dredge spoils reimplaced onsite in a 7-acre launch basin which also contains contaminated sediment. The dredge spoils will be placed behind an earth berm constructed to contain the sediments in the back of the launch basin. The entire launch basin, including the reimplaced sediments, will be capped with several feet of clean sediment and developed as approximately 6 acres of intertidal wetland. The new wetland will isolate the contaminated sediments from the environment and mitigate the 3 acres of onshore wetlands which will be lost during the soil remediation and site development.

The remediation plan was permitted by the Army Corps of Engineers under Nationwide Permit 38 for the cleanup of hazardous and toxic waste sites. All contaminated soil and sediments were retained on site.

CHAPTER 12

USING THIS BOOK EFFECTIVELY

12.1 SOCIETAL EXPECTATIONS

In the not-too-distant past, discovery of contamination at a site was equated with doom for the property. During pre-purchase environmental audits, even the suspicion that subsurface contamination might be present was enough to end a deal. If encountered during construction, contamination was a source of great confusion and concern. Tough construction superintendents and project managers, who got all their work done by the well-respected construction management techniques of "yelling and screaming," could be seen meekly listening to young geologists and bookish industrial hygienists. The first reaction at construction sites was to suspend all normal activity and turn over the operational control to health and safety officers.

Today the hazardous waste professional is expected to do his job quietly, efficiently, on-time and out of the way of actual construction. Experienced owners, reasonable regulators, and cost conscious construction managers are willing to cede only the contaminated corner of the site, and construction continues on the remainder of the site in a normal manner. Health and safety officers are expected to adhere to the same budget and schedule constraints as other task managers at the site. Many of the field and office tasks that were considered high technology or expert research have become routine matters that can be handled by experienced technicians and staff level professionals. In this atmosphere, it is important that everyone performing a task related to contaminated materials or hazardous waste delivers value. This chapter provides some guidance on how to use this book in three example settings – during phase I assessments, during construction at a contaminated site, and in case contamination is discovered in the groundwater at an existing property.

12.2 EXAMPLE 1: USE OF JUDGEMENT DURING SITE ASSESSMENT

Phase I assessment of an urban site in any industrialized region is likely to turn up some items of concern. Discovering the items of concern used to be expensive and an expert's purview, but now a modestly priced computerized list search can *rapidly* identify dozens of suspect and questionable items. Other items of concern can be ascertained by gleaning agency files, driving

in the neighborhood and by reviewing historical aerial photographs. The main skill now needed during pre-purchase site audits is how to dismiss chaff and identify the key piece of information that could be the deal-breaker. It is possible that a Phase II investigation may be needed to resolve the questionable item.

If the list search or any other source turns up an item of concern direct about the site itself – for instance, a past violation, history of disposal, or a report of a spill, then a phase II investigation is a must. If the site itself appears to be benign, then the influence of any adjacent sites takes on more importance. In assigning potential risk, distance to the site, the local geology and the type of spill or leak should be considered. If an adjacent property has or had leaky underground tanks in close proximity to the site, then some limited investigation along the affected boundary would be justified, even if the matter was resolved and the neighboring property has already been cleaned up. Any plume or leakage that entered the site under evaluation may still reside there.

When initiating a phase I audit, one is likely to end up with a wealth of information within a short time – much of it incomplete, and some of it contradictory. In resolving conflicts and identifying items for further investigation, both the severity and the frequency of violations should be considered. Aimless drilling and sampling rarely resolves any concerns and often muddies up the picture further. However, if there are genuine concerns, then a limited, focused Phase II investigation is justified. When in doubt, an investigator should err on the side of caution.

12.3 EXAMPLE 2: CONSTRUCTION AT CONTAMINATED SITES

In most urban centers, as many as one-third of the industrial and commercial construction sites have had some past industrial use. Many infrastructure projects such as highways, transit lines, tunnels or sewer lines traverse industrial and commercial centers and encounter land that has had past industrial activity. Public works, such as sewage treatment plants, ports, hospitals and schools, are often built for redevelopment. In such cases, presence of contamination at a construction site is the norm and not an exception. Brownfields are a special case where the contaminated site happens to be located in a commercially depressed area, and requires special tax considerations for redevelopment.

The main challenge during construction at a contaminated site is to maintain a civil engineering focus. The project team will need to recognize that an excavation is a necessity for supporting foundations and not a cleanup of hazardous wastes. Therefore, just the necessary amount of excavation should take place, even if contamination is encountered. Any contaminated soils remaining in the ground may need to be isolated from exposure by the use of impervious membranes or compacted fill, without increasing the volume of excavation. This approach will be the key to minimizing the remediation costs, as any contaminated soils that are excavated or disturbed will not be allowed for use as backfill, and will need to be treated or disposed of appropriately.

Construction of footings will typically involve shallow excavations, where the possibility of encountering contaminated soils is high. Drilling for any deep foundations, such as CIDH piles for support of heavy loads, including bridge abutments or high-rise buildings, will result in soil removal from depth, where the presence of contaminated soils is uncertain. Relocation of utilities will also necessitate some shallow excavations in potentially contaminated soils. Lowering of grade at a contaminated site may represent the largest quantity of soils excavated at one location. If the drilling for piles is done during the dry months, then groundwater is probably not going to be an issue. However, if work is performed during the wet months, contaminated water — most likely perched water —may be encountered even in shallow excavations and utility trenches.

Construction at a contaminated site requires four tasks related to the presence of contaminated soil, groundwater or hazardous waste, as described below.

Task 1: Site Characterization

If a previous characterization report includes borings that are located within the proposed excavation footprint, then additional fieldwork may not be necessary. Otherwise, several shallow hand-auger borings should be drilled for complete coverage of the areas to be excavated. Samples should be tested for contaminants of concern including hydrocarbons, volatile organic compounds, metals, PCBs, pesticides, etc. If groundwater is encountered, samples of the water should be taken and tested for the same contaminants.

Sampling for characterization of contaminants at depth can be performed in conjunction with geotechnical drilling and sampling.

Task 2: Excavation Plan

The results of site characterization should be used as input to the preparation of an Excavation Plan. The plan should include the following components:

Quantity and Distribution of Contaminants—A summary section should show the results of site characterization, including the estimated quantity and distribution of contaminated soils. If feasible, the concentration values should also be shown. Data should be presented in both tabular form and on proposed grading plans.

Excavation Approach—Depending on the contaminant distribution, an approach for performing the excavations should be devised. The shallow excavation for the abutments or utility trenches and the deep (drilled) excavations for CIDH piles should be discussed separately; drilling may require use of slurry if caving is encountered. Handling and disposal of any contaminated slurry or groundwater from the drilled piles should be addressed. Proposed equipment, stockpiling options, field logistics, hot-spot identification procedures, and site responsibilities should be defined. Contingency items, such as presence of unexpected contaminants or significantly larger quantities, should also be discussed.

Monitoring Plan—This should include instruments to be used, frequency of readings, locations for monitoring, decision criteria for handling measured concentrations and suggested thresholds and ceilings. General suggestions for controlling emissions should also be included.

Health and Safety Plan—This should include safe work levels, protective equipment, training requirements and contingency procedures for handling emergencies.

Permits—Suggested permits to be obtained by the owner, or to be maintained by the construction contractor, should be listed.

Task 3: Field Monitoring

When construction begins, field monitoring should be performed to maintain compliance with the approved excavation plan. Field personnel should be assigned for instrumental measurements, record keeping, and maintaining compliance. These records will be necessary to document appropriate procedures, to obtain closure, and to resolve complaints from the community, if any.

Task 4: Disposal and Closure Report

At the end of subsurface construction, all records, such as transportation and disposal manifests, permits, test results, chain-of-custody documentation and closure letters, should be compiled in a final report.

12.4 EXAMPLE 3: DISCOVERY OF CONTAMINATION IN GROUNDWATER

Discovery of contamination in groundwater beneath a site can be a distressing and perplexing matter to resolve. The contamination may be discovered at the site itself during a site investigation, during periodic sampling of groundwater monitoring wells, or reported at an adjacent site. In some cases, the site may be located atop a regional plume that originated miles away, and is migrating across the site. After some months or years, the plume may disappear due to natural groundwater flow or as a result of groundwater remediation performed by others outside the site boundary. In the intervening months or years the site may be branded as a contaminated site.

If contamination is discovered, reported, or inferred at a site, the first item to check is whether any drinking water aquifers are affected. Although all surface water and groundwater of a state may be protected under environmental laws, the implications of a contaminated aquifer are much more severe. In cases of shallow water-bearing zones or perched zones, the groundwater may be brackish or its quality may be otherwise degraded due to past natural and industrial causes. In such cases, any remediation needed to restore water quality may be minimal, and may just consist of removal of the source, such as a leaking underground tank.

If widespread contamination is present in the region, or if the contamination has clearly originated outside the site boundaries, it may be possible to obtain limited exclusions from regulatory agencies,such as the water boards, that will allow a sale or development to go ahead, pending the resolution of remedial options by responsible parties located beyond the site boundaries. Always more time consuming to resolve than questions of contaminated soils at a site, the resolution of groundwater contamination questions requires different and broader thinking.

12.5 CLOSURE: NEED FOR BALANCE

Environmental professionals dealing with contaminated soils and groundwater have a responsibility to suggest reasonable remedial schemes. Whether as a means of resolving property transfer questions, or in order to make a construction project go forward, the focus should be on cleaning up the environment and providing value to the society and the site owner.

APPENDIX

PHASE I ENVIRONMENTAL SITE ASSESSMENT SITE VISIT CHECKLIST

Job No.________________
Site Visit Personnel Name:________________
Firm:________________
Phone Number:________________
Date of Walkthrough:________________
Weather Conditions:________________
Name and Position of Site Contact:________________
Phone Number:________________
Present Owner:________________
Present Occupant:________________
Present User (if different from occupant):________________

1. GENERAL SITE INFORMATION

A. Property Name:________________

B. Property Location:________________
   (1) Mailing Address:________________
   ____
   (2) County:________________

C. Property Description (color photo of main facility, include date of construction and remodeling, construction materials, type of construction, general condition, and number of stories for building onsite):
   ____
   ____
   ____
   % of Site Covered by Paving:_____%; % by Buildings, Remaining Property is:________________
   (1) Property Size (square feet):________________
   Roads Adjoining the Property:
   North:________________
   South:________________
   East:________________
   West:________________
   Onsite Roads or Parking Facilities:________________
   ____
   (2) No. of Buildings:________________
   Building size (square feet):________________

D. Type of Heat/Cooling in Building:________________
   ____
   Type of Sewage Disposal System:________________

E. Present Site Activities (describe operations & procedures):
   (1)________________
   ____

F. Length of Current Operations:____________________

G. Hazardous Materials/Chemicals Used?____________________

(1) Raw material storage and handling (including solvents, paints, etc.). Describe quantity, form of material, and storage facilities (show chemical storage areas on Plot Plan) Take color photos:

(2) If product is hazardous, describe storage methods:

(3) Hazardous Materials Generated – include quantity, concentration, characteristics (ignitable, reactive, corrosive, EPA-toxic):

Chemical Storage Practices: Good__________ Average__________ Poor__________
Environmental Concerns (leaking or damaged drums, open containers, etc.):__________

(4) Hazardous material and/or hazardous waste disposal practices (drummed, sent to impoundment, recycled, treated onsite, disposed onsite) Take color photos:

(5) ____________________

(6) Hazardous waste disposal – Onsite or Offsite (if onsite, take color photos):
(a) If onsite, does facility accept waste from other sources?__________
(b) If offsite:
Quantity of hazardous waste generated per month:__________
Average time stored:__________
EPA annual report available?__________

H. Sanitary/waste water disposal method:__________
Any waste streams other than to sanitary sewer or municipal solid waste system?__________

I. Any vendors used to transport, treat, store, or dispose of waste?__________

J. Any landscaping or pool-type chemicals used?__________

K. Is Equipment or vehicle maintenance conducted onsite?__________

L. Above-ground Storage Tanks (take color photos of all ASTs onsite)
Size (gal.)__________ Age:__________
Materials:__________ Material Stored:__________
Condition:__________ Corrosion Protection Method:__________
Use of Stored Material:__________
Secondary Containment Facilities:__________
Visual Evidence of Leak/Spill:__________
Associated Underground Piping? (location, type, age, leaks):__________
Located onsite map:__________
Has the tank leaked in the past?__________
(If yes, record in Section V, Spill Incidents.)

M. Underground Storage Tanks (USTs) Take color photos of location and associated vent/fill pipes
Capacity (gal.)________________ Age:________________
Construction materials:________________ Material Stored:________________
Condition:________________ Leak Detection:________________
Corrosion Protection Method:________________
EPA/State Registered?________________
Permit or Registration Number:________________
Leak Test Report (date):________________

Visual Evidence of Leak/Spill:________________
Use of Stored Material:________________
Associated Underground Piping? (Location, type, age, leaks):________________
____
Located on site map:________________
Has the tank leaked in the past?________________
(If yes, record in Section 5, Spill Incidents.)

N. Recycling Units? (Locate on Plot Plan.)
Possible PCB-Containing Transformers & Equipment________________

O. Transformers (photo, locate on Plot Plan)________________
(1) How many (locate on Plot Plan)________ Ownership?________ ID#________
(2) Type (i.e., pad or pole-mounted)?________________
(3) Records relating to PCB items:________________
(4) Inspect Area(s) for Spill/Leakage:________________
(5) Any Reported Spills?________________
(6) Non-PCB label present on transformer?________________
Hydraulic Equipment:________________
Type:________________
Location:________________
Evidence of Leakage?________________
PCB-Containing Hydraulic Fluid? Yes________ No________ Unknown________

P. Pits and Sumps (color photo, location on Plot Plan)
(1) Location:________________
(2) Nature of Use:________________
(3) Type of Material Handled:________________
(4) Location or Site Map:________________

Q. Recent Spill Incidents (onsite) (show location on Plot Plan)
(1) Substance spilled:________________ Quantity (gal.):________________
Date:____/____/____ Location:________________
Remedial Action:________________
Name and Address of Facility where contaminated soils or adsorb pads, etc. disposed of:
____
____
Remaining Contamination:________________
____

R. Other Known Soil or Water Contamination:________________

S. Groundwater Monitoring Programs
(1) Name of Program Manager________________
Type of Substances Monitored:________________
Current Status:________________

2. SITE OBSERVATIONS (WITH COLOR PHOTOS)

Limiting conditions (locked doors, inaccessible areas, adjacent buildings, asphalt or paved areas covering spill sites, bodies of water, snow, rain):________________

____
____
____

A. Surface Water Discharge(s)
Location:______________________________
Condition:______________________________
Permit Type:______________________________

B. Distressed Vegetation Areas
Location:______________________________
Area (sq. ft.)______________________________

C. Any Evidence of Solid Waste Dumping or Disposal?______________________________

D. Discolored Ponds or Flowing Waters?______________________________

E. Onsite Wells (locate on Plot Plan):
Springs, Cisterns Onsite?______________________________
Type:______________________________

Status (in use, dry, not used)______________________________
Date of Construction:______________________________
Well Logs Available:______________________________
Is Well Known to be Contaminated?______________________________
If Yes, List Type and Source of Contamination:______________________________
Groundwater Wells, Springs, or Cisterns onsite?______________________________
Well Depth (ft.)______________________________

F. Cesspools, Septic Tanks:
Location:______________________________
____
Status:______________________________

G. Abnormal Odors (gasoline, oil, etc.)______________________________

H. Discolored Surface? Staining?______________________________

I. Recent Soil disturbances:______________________________
Fill Added:______________________________
Source of Fill:______________________________
Evidence of Dumping Hazardous Material, Debris, or Construction Material (photo):______________________________
____

J. Underground Utilities and Piping
(1) Location of Underground Utilities (if available):______________________________
(2) Location of Storm Drains and Sanitary Sewers:______________________________

K. Hydrogeology
(1) Direction and Quantity of Surface Water Discharge:______________________________
(2) Depth of Groundwater (ft):______________________________
(3) Probable Groundwater Flow Direction:______________________________
(4) Visual Evidence of Run-on/Runoff: Yes____________ No____________
(5) Visual signs of Contaminated Run-on/Runoff: Yes____________ No____________
(6) Nearest Surface Water Body:______________________________
(7) Source(s) of water for residents or other establishments in area:______________________________
Topographic Conditions
Site:______________________________
Surrounding Area:______________________________

3. **PAST LAND USE**

A. Previous Owners (give dates):____

B. Previous Occupants or Users (give dates):__________________________________

C. Previous Site Activities (give dates):____________________________________

D. Previous Spills, Releases, or Fires? (give dates, quantity, and location):________________

(1) Substance Spilled:________________ Quantity (gal.)______________

Date:____/____/_____ Location:____________________________

Remaining Contamination:____

Other Known Soil or Water Contamination:__________________________________

4. **SURROUNDING AREA**

A. Adjacent Facilities

(1) Descriptions:

North:____

East:____

South:____

West:____

B. Environmental concerns at Adjacent Properties

Look for and ask client contact if adjacent properties use or store chemicals, i.e., industrial establishments, service stations, dry cleaners, etc. Types of chemical used or stored? How much? Are wastes handled and disposed onsite? History of spills or leaks (describe below)? Evidence of staining? Sheen? USTs or ASTs present?

(1) Descriptions:

North:____

East:____

South:____

West:____

C. Spill Incidents (adjacent sites)
   (1) Substance Spilled:________________ Quantity (gal.):________________
   Date:_____/_____/_____ Location:____
   Remedial Action:____
   ____

D. Other adverse environmental conditions/past enforcement actions at adjacent sites:________________
____
____

E. Types of businesses surrounding the site within a 1/4 mile radius (note number of gas stations, dry cleaners, etc. and their distance and direction from the site):________________________________
____
____

5. **ZONING**
   A. Zoning at the site:____

   B. Description of allowed uses:_______________________________________
   ____
   ____

GLOSSARY

accuracy. Nearness to the true value.

ACMs. Asbestos Containing Materials. Any material containing 1% or more asbestos.

action levels. Contaminant concentration levels set by a regulatory agency that trigger action such as remediation or removal.

adsorption. Adherence of contaminants to the surfaces of soil particles, as compared to absorption which involves contaminant entry into particle pores.

advection. Movement of contaminants in a fluid or gas due to temperature changes.

aerobic condition. A condition during bioremediation when air and oxygen are available to the bacteria.

AHERA. Asbestos Hazard Emergency Response Act of 1986.

air permeability. The rate at which air can pass through a porous medium.

air sparging. Injection of air into the subsurface below the water table in order to cause volatilization of contaminants into the vadose zone.

air stripping. Passing water through a column in which stirring action separates the dissolved volatile contaminants from the water. This can also be accomplished in waterfalls, fountains or cascades.

alluvial soils. Soils derived from the action of streams or rivers.

analytes. The contaminants that are the target of analysis during testing.

anthracine. One of the polycyclic aromatic hydrocarbons that is found in manufactured gas (towne gas) wastes.

aquifer. A sandy soil layer that can store groundwater, useful for water supply or irrigation using wells.

ARARs. Applicable or Relevant and Appropriate Requirements.

aromatic volatile organics. A group of organic compounds that have distinct aroma and are volatile. Generally, these include a benzine ring in their molecule.

ASTM. American Society for Testing and Materials.

atomic absorption. A property of elements used in analytical testing for detecting small concentrations.

bailer. A device used to obtain water samples from a watering well.

baghouse. A device made of canvas bag, used for collecting particulates from the off gases, works similar to a domestic vacuum cleaner bag.

bedding. Layering in soils or rock, or the boundary between two layers.

bell curve. Representation of a normal distribution of a population. Soil contaminant levels that show a typical bell-shaped curve indicate random sampling would provide representative samples.

bench scale study. A test study in a lab where a small quantity of contaminated material is tested in a very precise manner, so the results can be reliably extrapolated to a large-scale test.

Benzo(a) Pyrene. One of the polycyclic aromatic hydrocarbons that is highly carcinogenic and is generally used in health risk studies as a benchmark for the total polycyclic aromatic hydrocarbons.

biocell. A stockpile of contaminated soil undergoing bioremediation under a synthetic liner cover. It incorporates piping for injecting air, adding or nutrients, and for extracting vapors.

biodegration. The natural degradation of contaminants due to the action of microbes which changes toxic materials to less toxic forms.

biomass accumulation. One of the remediation techniques in which plant materials such as trees, plants, etc. are allowed to preferentially uptake or accumulate contaminants from the soil.

biopile. See biocell.

bioremediation. Treatment techniques that encourage the growth of microbes such as misophiles that consume hydrocarbons and other contaminants.

bioventing. A method in which volatile contaminants are allowed to volatilize slowly so that they can be consumed by microbes, aided by the availability of oxygen through venting.

blow counts. The number of blows it takes to drive a sampler into a soil using a standard hammer weight. This is an indicator of the resistance of soil to driving or its density and shear strength. See SPT.

boring log. A record of the various soil types, layers, and strata that may be encountered during drilling and sampling, usually prepared in the form of a graphic table. Vertical columns show the description of soils encountered, blow counts, and measurement of any volatiles.

CAA. Clean Air Act.

cap. A cover placed over contaminated area in order to reduce infiltration of water and to reduce any volatilization into the air. In its simple form could consist of a paved area, and usually incorporates synthetic liners.

capillary fringe. The zone immediately above the water table in which water rises due to the capillary action. The height of the capillary fringe depends on the soil type and microfabric.

carbon adsorption. The adherence of volatile contaminants to the surfaces of charcoal or granular activated carbon.

carbon dioxide uptake. One of the parameters measured during bench scale study of bioremediation, measuring how much carbon dioxide is being emitted by microbes, a measure of the remediation effectiveness.

catalytic oxidation. A method of treatment for volatile contaminants. A common example is a catalytic converter in a car which oxidizes hydrocarbons at high temperatures.

CFUs. Colony forming units, a measure of the number of bacteria present in a unit quantity of soil.

chain of custody. A record of sample possession from the time a sample is obtained in the field, documenting the transport, storage, arrival , and testing at a laboratory. This document is important in ascertaining the safekeeping of the sample and the veracity of the test results.

chemical of concern. Any contaminant, which due to its high concentration, or high toxicity, becomes the primary contaminant that will determine the outcome of a site.

chemically stabilized waste. Waste materials, such as contaminated soils, that have been rendered inert by mixing with pozzolins, epoxies, or other chemicals.

chlorinated VOCs. Volatile organic compounds that contain one or more molecules of chlorine. See halogenated volatiles.

clay matrix. A clayey soil deposit or excavated clay soil.

CIDH piles. "Cast-In-Drilled-Holes" piles. Deep foundation constructed by using an in-place construction technique that involves drilling a hole, inserting reinforcing steel, and pouring concrete.

clean-up criteria. Chemical concentrations that may be prescribed by a regulatory agency as targets to be achieved at the end of remediation.

coarse-grained fraction. Sand and gravel sized particles in a soil mass. Generally, this fraction has less surface area and thus less contaminant adhering to the surface.

cohesionless soil. *See* **coarse grained fraction**.

cohesive soil. *See* **clay matrix**.

compaction. Method for densifying the soil or other inert waste material after treatment for use as backfill.

computer modeling. Use of computer program to simulate the behavior of groundwater or movement of dissolved chemicals, used in contaminant fate and transport studies.

condenser. Post-treatment device used to facilitate the sedimentation of suspended particulates.

confidence level. A statistical measure of the certainty with which the presence of contaminant at a site is known.

confirmation sampling. Sampling performed at the end of remediation in order to ascertain if the site has been adequately remediated.

containment. A method of remediation in which the impacted zone is confined by building barriers and caps, thus preventing the contaminant from infiltrating the groundwater or the atmosphere.

contaminant transport. The movement of contaminants away from the source, studied in conjunction with the groundwater studies.

CPT. Cone Penetrometer Test: a truck-mounted device for probing the ground.

cradle to grave. Environmental Protection Agency's authority for controlling chemicals, beginning from their manufacture, and including transport, spillage, recovery, treatment, and eventual destruction or disposal.

cyclone. A post-treatment device used to separate the suspended particulate fractions from off gases.

CWA. Clean Water Act.

decontamination facility. Facility used to remove waste materials adhering to a worker's clothing, construction equipment, or trucks.

degree of freedom. A statistical measure of the shape of a bell curve.

deminimus level. Very low levels of contaminants that may be allowed to remain on the site, similar to background levels in the area.

depositional environment. The environment in which a sediment may have originally been deposited; for example glacial, alluvial, aeolian, etc.

dermal exposure. Exposure of human skin to certain contaminants that have the ability to enter the tissue and bloodstream through the skin.

desorber. Device such as a kiln used to heat impacted materials so the volatile contaminants would leave the surface, or desorb.

detection limits. A threshold concentration during analytical testing, below which the test results below which the test results are not reliable. This limit depends on individual chemicals, their concentrations, presence of other chemicals and the test method.

deterministic criteria. Clean-up criteria that may be based on a specific target concentration of chemicals (rather than on health risk).

dig and haul. Excavation and off-site transport of impacted materials.

directional drilling. A drilling method adapted from oil drilling techniques, used for drilling wells at low angles to reach inaccessible areas for in-situ remediation.

DNAPL. Dense Nonaqueous Phase Liquids – immiscible liquids that are denser than water and would sink to the base of an aquifer, e.g., TCE, PCE.

draw down rate. A measure of the rate at which water can be withdrawn from a well during pump tests.

effluent. Liquids or fluids that may emanate from an impacted area or installation.

EM wand. A geophysical device that is used for detecting presence of subsurface metallic object or objects of different conductivity using electromagnetic signals.

encapsulation. A technique for stabilizing contaminated materials by adding a chemical that encapsulates or surrounds the contaminated soil particles.

EPA. Environmental Protection Agency. The agency responsible for regulating hazardous wastes, directly or by delegating some authority to the states.

exposure pathways. Pathways along which contaminants travel from the source to sensitive receptors.

extraction methods. Methods used in the laboratory for separating the contaminants from the soil matrix, in preparation for analysis.

environmental fate and transport modeling. Predictive methods for estimating the spread and concentration of chemicals at some distance away from the source of contamination, used in health risk assessments.

Environmental Site Assessment (ESA). A preliminary review of the history and condition of a site, intended to assess impact from contamination, disposal, etc. Also termed Preliminary Site Assessment or Phase I Assessment.

EOR. Enhanced Oil Recovery. A recovery method used in oilfield production involving injection of steam into wells, adapted for remediation of sites impacted with viscous, semivolatile compounds.

epidemiological studies. Studies in which health effects of low levels of contaminants on large human populations are assessed.

fabric. The arrangement of grains in a soil mass.

Feasibility Study (FS). A study to compare methods of remediation and select the best alternative

FID. Flame Ionization Detector. A hand-held instrument for detecting the presence of volatile organic compounds.

field test kits. Portable test kits to measure the concentrations of various chemicals by using comparative methods.

FIFRA. Federal Insecticide, Fungicide and Rodenticide Act of 1972.

fine-grained fraction. Clay and silt sized particles, generally flat-shaped, cumulatively representing a large surface area to which contaminants can adhere.

fluidized beds. The use of air pumped upwards through a granular mass to suspend particles for better blending during treatment, a technique adapted from coal-burning furnaces.

focused sampling. Sampling of hot spots performed by a person knowledgeable about the site Also called authoritative sampling or biased sampling.

fugitive dust. Dust escaping a site during excavation or other remediation.

GAC (Granular Activated Carbon). Charcoal granules or other carbon in which the porosity has been enhanced, used to capture volatiles in off-gas treatment.

gas chromatography. An analytical laboratory technique used to identify chemicals by using their characteristic property of absorbing selective wavelengths of light in vapor or gaseous form.

generator. Any person whose act produces a hazardous waste.

geologic origin. Identification of the processes which formed a rock (volcanic, sedimentary, etc.) or soil (alluvial, glacial, etc.)

geologic setting. A description of the significant geologic features surrounding a site location (e.g., a flood plain, adeltaic fan, a sedimentary basin, etc.)

geostatistical methods. Methods of assembling and analyzing sample test results based on the application of statistical techniques to the spatial distribution of chemical concentrations.

GIS. Geographic Information Systems. A group of computer data base methods that record and present information with reference to physical location of sampling points.

gradational analysis. Measurement of grain size distribution of soil or other particulate matter.

granular. Refers to sand and gravel size soils. *See* **coarse-grained fraction**

ground penetrating radar. An instrument for detecting the presence of buried objects, using a radar source and receiver.

groundwater. Water contained in aquifers or other soil and rock layers below the ground surface.

grout curtain. A subsurface barrier created by injecting grout along a row of points.

halogenated semivolatiles. Organic compounds with mild volatility that contain one or more halogen atom in their molecules.

hazard index. A measure of the health risk caused by a noncarcinogen.

Hazard Ranking System. A method of assigning risk or hazard to candidate site, including a point system for size of spill, toxicity of contaminants, proximity to aquifers or to populations.

hazardous waste. A waste material or product that is "listed" as a RCRA waste, or has hazardous characteristics—ignitability, corrosivity, reactivity or toxicity.

HDPE. High Density Polyethylene. A sturdy plastic sheet used as a liner in pits, ponds, landfills, etc., resistant to most chemicals.

heavy hydrocarbons. Hydrocarbon with long chain molecules, generally C15 and above, not very volatile and have low mobility.

Henry's Law. A chemical law governing the relationship of vapors and liquids in proximity to each other, used in determining the viability of vapor extraction for contaminants.

holding time. Time allowed for testing of a contaminant after sampling. Beyond this time degradation occurs and the test results are not reliable. Seven days for most volatile organic compounds.

hot spots. Areas of past disposal, spillage, or leakage, with high concentrations of contaminants.

ICP. Inductively coupled plasma, a laboratory technique used in analytical testing.

ID. Internal diameter.

IDLH. Immediate Danger to Life and Health, denoting the chemical concentration that could lead to fatal results without respiratory protection, allowing 30 minutes to escape the area.

immiscible. A liquid that does not mix with water, such as oil.

ingestion. Eating or drinking of chemical directly into the body.

inhalation. Introduction of contaminant into the body by breathing.

innocent buyer. A buyer who has performed an environmental assessment and other due diligence steps before purchase.

inorganic wastes. Chemical wastes that do not include carbon molecules and are not derived from plants or hydrocarbons, e.g., nitrate or asbestos.

in-situ. In place, refers to remediation performed without excavation.

intake rate. The rate at which humans may inhale or ingest chemical of concern, used in health risk studies.

intrusive techniques. Sampling or testing techniques that require drilling, excavation, or penetration of the contaminated area.

isolation. Remediation technique in which the contaminated zone is prevented from entering the environment, without any reduction in toxicity or volume.

lagging. Wooden planks placed between soldier piles (I-beams) during shoring of excavations.

lamp black. A powdery black waste material derived from PAHs found at town gas sites. It contains carcinogenic materials.

land ban. The policy of eliminating land disposal of bulk liquids and dioxins, and limiting the disposal of other hazardous wastes without treatment.

lead agency. A regulatory agency that takes the lead in overseeing investigation or remediation at a site, provides a sign-off or approval at the end, and acts as the intermediary in dealing with other agencies.

leach. Removal of contaminant by seeping out of a mass, induced in order to collect the contaminant.

limited access rigs. Drill rigs used for indoor or confined area drilling and sampling.

LNAPL. Light Density Nonaqueous Phase Liquids. Immiscible materials that float on water such as gasoline.

local geology. The types and arrangements of soil and rock in the immediate vicinity of the contaminated area.

LUST. Leaking Underground Storage Tank.

mass spectroscopy. A technique used for analytical testing.

MEK. Methyl Ethyl Keytone. One of the constituents of solvents that has been found at very widespread sites.

MTBE. Methyl Tertiary Butyl Ether. A constituent of clean burning oxygenated gasolines that has become persistent contaminant in groundwater.

mesophiles. Types of microbes that consume hydrocarbons.

micro tip. A hand-held device for detecting the presence of volatile hydrocarbons.

microfabric. *See* **fabric**.

miscible. A liquid that easily mixes with or dissolves into water.

mitigative action. An action taken in order to remediate or remove a contaminated material.

models. See computer models.

moisture content. The amount of moisture contained in a sample of soil, measured in percentage by weight.

monitoring well. A well installed in order to check the water levels, and obtain samples of groundwater for testing.

NAPL. Nonaqueous Phase Liquid. A liquid that does not easily mix with water.

ND. Nondetectable. Generally a contaminant at a concentration below the detection limit.

NESHAPs. National Emission Standards for Hazardous Air Pollutants Program.

NIOSH. National Institute of Safety and Health.

No. 2 Heating Oil. Type of fuel oil used in residential and small commercial establishments for winter heat and stored in an underground tank.

nonhalogenated VOC. Volatile organic compounds that do not contain any atoms of chlorine or other halogens.

microtunnelling. A technique used for installing a small diameter pipe as an aid during in-situ remediation.

nonintrusive techniques. Testing techniques that do not require drilling or penetration of the subsurface. This would include all geophysical techniques.

NPDES. National Pollution Discharge Elimination System.

NPL (National Priority list). A list of sites that are intended to be remediated under Superfund.

normal distribution. Distribution of contaminant concentrations at a site that would plot as a bell curve.

normal population. A population of samples at a site that follows the normal distribution.

nutrient levels. The concentration of certain basic chemicals that are needed by microbes in order to reproduce and survive at a site.

N-value. *See* **blow counts**.

OSHA. Office of Safety and Health Administration.

oxygen intake. One of the measures of bioremediation during bench scale studies showing the amount of oxygen being consumed by bacteria; an indicator of hydrocarbon consumed.

packing column. A device used in the laboratory for separating contaminants.

PAH. Polycyclic Aromatic Hydrocarbons; also referred to as PNA or polynuclear aromatics, byproducts of towne gas sites. These are generally heavier hydrocarbons, yet they have an aroma, such as camphor, which is one of the PAHs. Most of the are carcinogenic.

particulate emissions. Emissions of contaminants from a site in a particulate forms that are suspended in air.

particulates. Small particles of contaminated materials.

PCB. Polychlorinated Biphenyles.

PCE. Tetra Chloro Ethylene, formerly referred to as Per Chloro Ethylene when it was considered to be the fully saturated form.

PEL. Permissible Exposure Limit. One of the measures of contaminant levels referred to as a health risk precaution.

perched groundwater. Groundwater that is not widely distributed, and has unsaturated zones below it.

petroleum exclusion. The exclusion from RCRA regulations given to petroleum products and crude oil in the original form. However, if the petroleum products are leaked or spilled, they are dealt with as waste.

Phase I Assessment. An initial environmental assessment that is done by inspecting the site and looking at historical data, but without any sampling.

Phase II Investigation. An intrusive, detailed investigation of a site including drilling, sampling, and laboratory testing.

photo ionization. One of the detection methods used in hand-held instruments.

photo VAC. One of the hand-held instruments for detecting volatile organics.

plasticity of soils. A property of soil that indicates their behavior when mixed with water.

plot plan. A map showing the location of equipment, buildings, etc. at a site.

plume boundaries. Boundaries of an area that has been contaminated.

PM10. Particulate matter finer than 10 microns --the breathable dust range. These are often sampled and tested separately, because they pose a higher health risk. than the total dust.

penetration resistance. Resistance of a soil layer to penetration by a sampler, expressed as blow counts.

portable detection instruments. See hand-held instruments.

POTW. Publicly Owned Treatment Works, or sewage treatment plant.

PPM. Parts per million, indicating the level of contaminants. More accurately described as milligrams per kilogram for soils and milligrams per liter for groundwater.

precision. The repeatability of a test result.

predrainage. Dewatering technique in which wells are installed and the ground water lowered before excavation.

Primavera. A computer program for tracking the progress of a project. It includes a number of project control items including showing schedule, etc.

Pedology. The study of the development of shallow soil horizons.

project controls. Schedule, budget, resource loading, cash flow, etc., the criteria that are used to monitor a project.

PRPs. Potentially Responsible Parties. Property owners or others who may potentially be required to finance cleanup of a site.

pump and treat. A group of treatment methods in which contaminated groundwater is pumped out and then treated above ground.

pumping test. A test in which water is withdrawn from a well in order to test the aquifer by measuring well discharge.

purge and trap. Used in the laboratory during analytical testing as an aid for extracting or separating the contaminant from soil.

QA/QC. Quality Assurance and Quality Control.

radius of influence. The area around a well from which it can extract contaminants used for vapor extraction (air wells), and pump and treat (waterwells).

Radon. A colorless odorless radioactive gas that is formed by the decay of medium.

RAGS. Remedial Action Guidance for Superfund; EPA document that defines health risk analysis methodology.

random numbers. A set of numbers in which every number has equal chance of being selected.

random sampling. A sampling method in which every point at a site has an equal chance of being sampled. A technique for obtaining representative samples at site with normal distribution of contaminant levels.

randomly selected. Selected in a manner that any of a group of items has a an equal chance of being selected.

RCRA. Resource Conservation and Recovery Act of 1976, giving EPA the authority to regulate hazardous wastes from cradle to grave.

RCRA closure. A site closure performed in accordance with RCRA guidelines.

reference dose. A certain contaminant concentration that could be consumed without serious effects on a regular basis, used in health risk assessment of noncarcinogenic contaminants.

regulatory threshold. Chemical concentration beyond which regulatory action would be triggered.

remediation. A method, process, or technology that can achieve reduction in volume, toxicity, or mobility or a contaminant.

remote sensing. Methods of surveying or observing a site from a distance, such as aerial photographs.

resistivity techniques. One of the geophysical methods that depends on the electrical conductivity of near -surface soils.

RI/FS. Remedial Investigation and Feasibility Study. Refers to the formalized methodology for investigation and selection of remedial methods used in Superfund investigations.

risk-based action levels. Contaminant concentrations that are derived on the basis of health risk to humans.

ROD. Record of Decision. A legal document binding the site owner and the lead agency for remediation of a site.

SARA. Superfund Amendments and Reauthorization Act of 1986 that amended the original Superfund authorization of 1980.

SCBA. Self-Contained Breathing Apparatus, used for entry into heavily contaminated sites.

scrubber. A device for removing particulate matter from air.

sediments. Soils or other mineral materials that have been derived as a result of alluvial action.

segregated. Separated by contaminant type or physical characteristics.

semivolatiles. Hydrocarbons or other organic compounds that have large molecules, are also mildly volatile, many have a characteristic aroma or odor, and have significant amount of residue and limited mobility.

shear strength. The strength of a soil such as sand or clay that determines a stable slope angle for excavation sides.

sheet piles. Steel sheets that are driven into the ground as a barrier to reduce intrusion of water and to support the sides of an excavation.

simple random sampling. Random sampling that is dependent only on one variable, such as one type of contaminant, or at one depth in an excavation.

slant boring. Borings drilled in a straight line that is at an angle from the vertical. This is different from directional drilling where the direction of drilling can be changed and a well can be drilled along a curved path.

slant wells. Wells installed in a slant boring.

slope factor. A measure of the carcinogenic toxicity of a material, used in health risk analysis.

slug test. A type of test for a well to determine characteristic of the aquifer done by inserting a plunger or a slug down the well, thus depressing the level and then observing as the water level rises back.

soil boring. A boring drilled to sample soils.

soil classification system. Methods of organizing soil types by their physical characteristics. An example is the Unified Soil Classification System (USCS) which depends on grain size and plasticity.

soil gas. Air or vapors contained within the pores of soils below the ground surface.

soil matrix. The mineral matter, in the subsurface, aside from water or any contaminants.

soil pores. The space between grains of soil.

soil sampling. Sampling of soils, generally done by driving a sampler.

soil washing. Use of detergents or solvents to separate the contaminants from a soil, also soil flushing.

site characterization. *See* **Phase II investigation.**

soldier piles. Vertical steel I-beams used to support the sides of an excavation.

solidification. A method of stabilization in which the contaminated soil is mixed with a cementing agent, thus rendering the material solid.

solvents. Organic chemicals that are nonaqueous and can dissolve oily substances; generally used to clean machine parts and electronic components.

spatial distribution. Distribution of contaminants at a site over a geographic area.

spectroscopy. A laboratory method used in the analytical testing of contaminants.

stabilization. A group of remediation methods that include encapsulation, solidification, and fixation, using cementing agents to reduce the mobility of contaminants.

Standard Penetration Test. Test performed to measure the penetration resistance of a standard sampler (2-in. diameter) by dropping a standard hammer (140 lb.) through a standard height (30 in.).

stockpiles. A pile of contaminated material separated as a temporary storage after excavation, awaiting characterization and final disposition.

strata boundary. Layer boundaries between adjacent beds or strata of soil or rock.

stratigraphy. The description of layering of soil and rock at a site.

strippability. The property that is indicative of the ability to separate a contaminant from the matrix or soil grains.

Student's "t" method. A method used to measure the confidence level during a statistical evaluation of confirmation test results.

supplied-air respirator. Respirator with a hose that is used to supply breathable air to someone working at a highly contaminated site.

surfactant flushing. Use of a detergent or surfactant to reduce the surface tension, and pump water to the subsurface to flush out contaminant from the soil.

Surtrak. A computer program that is used to track project controls.

synthetic liner. A plastic liner made of polyethylene or other materials used as a barrier or cap to contain or limit the movement of a contaminated material.

tank invert. The lowest point of an underground storage tank.

TCE. Trichloroethylene, one of the common components of solvents.

test pit. A small excavation performed to obtain samples of soils and contaminants.

thermal desorption. Use of mild heat to volatilize contaminants off from soil grains.

TLV. Threshold Limit Value, a measure of the chemical concentration that may require health protection.

total suspended particulate. The total load of granular or particulate material that is suspended in an emission including both breathable and nonbreathable dust.

toxicity. The property of a chemical to affect living things in a negative manner.

treatability studies. Studies done in the lab at a small scale to apply different methods of treatment to contaminated materials.

TSD. Transfer, Storage or Disposal facility such as a landfill or recycler.

TWA. Time-weighted average, a measure of permissible amount of contaminant levels.

Unified Soil Classification System. A method of classifying soils based on their grain size and plasticity.

unsaturated zone. *See* **vadose zone**.

unsorted. Well-graded soil that contains various grain sizes.

UST. Underground Storage Tank.

vadose zone. The unsaturated zone between the water table and the ground surface.

vapor extraction. Method of removing contaminated volatile materials from the vadose zone by using wells and vacuum pumps.

vapor phase chromatography. A laboratory technique used to identify contaminants in the laboratory.

viscosity. A property of contaminants that tells us about their ability to flow.

VOCs. Volatile Organic Compounds. A group of organic compounds that volatilize easily, have low boiling point, can easily be breathed, and affect health of living organisms.

water table. The surface of groundwater in the subsurface. The depth to water table is generally one of the items investigated and recorded at a site during an investigation.

wilting point. The moisture content in soil at which plants will begin to wilt, a measure of moisture content.

BIBLIOGRAPHY

The following bibliography contains recent references and suggested reading material.

Acar, Y.B. and Daniel, D.E.; Editors (1995). *Geoenvironment 2000: Characterization, Containment, Remediation and Performance in Environmental Geotechnics*. ASCE Geotechnical and Environmental Engineering staffs, *Geotechnical Special Publications: No. 46*; Available from the American Society of Civil Engineers, Reston, Virginia, 2 vols., 1822 pp.

Alleman, B.C. and Leeson, A. (1997). *In-Situ and On-Site Bioremediation*. Papers from the Fourth International In-Situ and On-Site Bioremediation Symposium, New Orleans, Louisiana, April 28 - May 1, 1997. Battelle Press, Columbus, Ohio.

Allen, J. (1997). *Annual Editions: Environment 97/98, 16th ed.* Brown & Benchmark, Madison, Wisconsin, 256 pp.

American Society for Testing and Materials, (1995). *ASTM Standards on Environmental Site Assessments, etc.* ASTM, Philadelphia, Pennsylvania, 62 pp.

American Society for Testing and Materials, (1997). *ASTM Standards Related to Environmental Site Characterization*. ASTM Committee D-18 on Soil and Rock, Philadelphia, Pennsylvania.

American Society of Civil Engineers, (1996). *Environmental Site Investigation Guidance Manual*. ASCE Manual & Report on Engineering Practice, Vol. 83, Reston, Virginia, 160 pp.

Anderson, W.C.; Editor (1993). *Innovative Site Remediation Technology,* Vol. 6: Thermal Desorption. American Academy of Environmental Engineers, Annapolis, Maryland, 120 pp.

Anderson, W.C.; Editor (1994). *Innovative Site Remediation Technology,* Vol. 4; Stabilization/ Solidification. American Academy of Environmental Engineers, Annapolis, Maryland.

Anderson, W.C.; Editor (1995). *Innovative Site RemediationTechnology,* Vol. 3: Soil Washing/Soil Flushing. Springer Verlag, Heidelberg, Germany, 174 pp.

Bajpai, R. and Zappi, M.; Editors (1997). *Bioremediation of Surface and Substance Contamination*. Annals of the New York Academy of Sciences, Vol. 830.

Baker, K.H., et al. (1994). *Bioremediation*. McGraw-Hill Text, New York, New York, 375 pp.

Barcelona, M., Gibb, J., Helfrich, J. and Garske, E., (1988). *Practical Guide for Groundwater Sampling*. St. Mut. Press, United Kingdom, 156 pp.

Bedient, P.B. (1994). *Ground Water Contamination, Transport and Remediation*. Prentice Hall, Upper Saddle River, New Jersey, 560 pp.

Boulding, J.R.; Editor (1996). *U.S. Environmental Protection Agency Engineering Sourcebook.* Ann Arbor Press, Inc., Ann Arbor, Michigan.

Buonicore, A.J., (1995). *Software Package in Support of ASTM Standard Practice for Environmental Site Assessments, etc.* ASTM, Philadelphia, Pennsylvania, 54 pp.

Chaudhry, G.R. (1995). *Biological Degradation and Bioremediation of Toxic Chemicals.* Timber Press, Inc., Portland, Oregon, 520 pp.

Cheremisinoff, P. (1992). *Guide to Underground Storage Tanks: Evaluation, Site Assessment, and Remediation.* Prentice Hall, Upper Saddle River, New Jersey, 367 pp.

Chrisman, N.R. (1997). *Exploring Geographic Information Systems.* John Wiley & Sons, Inc., New York, New York, 320 pp.

Coates, D. (1981). *Environmental Geology.* John Wiley & Sons, Inc., New York, New York, 731 pp.

Collert, J.; Editor (1997). *Environmental Health and Safety CFR Training Requirements, 3rd ed.* Government Institutes, Inc., Rockville, Maryland, 450 pp.

Cooper, A. (1996). *ESA's Made Easy: A Checklist Approach to Phase I Environmental Site Assessments.* Government Institutes, Inc., Rockville, Maryland, 206 pp.

Corporate Law and Practice Course Handbook, 1985-86 Ser., Vol. 797, (1992). *The Impact of Environmental Regulations on Business Transactions and Operations 1992.* Practicing Law Institute, New York, New York, 810 pp.

Cranor, C.F. (1997). *Regulating Toxic Substances: A Philosophy of Science and the Law.* Oxford University Press, Inc., New York, New York, 272 pp.

DeLong, J.V. (1997). *Property Matters.* S&S Enterprises, Los Angeles, Calif., 352 pp.

Demers, M.N. (1996). *Fundamentals of Geographic Information Systems.* John Wiley & Sons, Inc., New York, New York, 480 pp.

Dupont, R.R., et al. (1998). *Bioremediation: Design and Application.* American Academy of Environmental Engineers, Annapolis, Maryland, 450 pp.

Evans, J.C.; Editor (1997). *Proceedings of the Conference on In Situ Remediation of the Geoenvironment.* American Society of Civil Engineers 1997 Annual Convention; Minneapolis, Minnesota, October 5 - 8, 1997. Available from the ASCE, Reston, Virginia.

Flathman, P.E., et al.; Editors (1993). *Bioremediation: Field Experience.* Lewis Publishers, Inc., Boca Raton, Florida 560 pp.

Freeman, H. and Harris, E.; Editors (1995). *Hazardous Waste Remediation: Innovative Treatment Technologies.* Technomic Publishing Company, Lancaster, Pennsylvania, 361 pp.

Gee, G.W. and Wing, N.R.; Editors (1994). *In-Situ Remediation: Scientific Basis for Current and Future Technologies*. Thirty-third Hanford Symposium on Health and the Environment, November 7 -11, 1994, Pasco, Washington. Battelle Press, Columbus, Ohio, 2 vols., 1272 pp.

Geltman, E.G. (1997). *Environmental Audits: Self Policing for Environmental Protection*. American Bar Association, Chicago, Illinois.

Gill, M. and Kusiner, J., Jr., (1994). *Proceedings of the 15th Annual Meeting of the Society of Environmental Toxicology and Chemistry.* Denver, Colorado, October 30 - November 3, 1994. Available from SETAC Press, Pensacola, Florida.

Guruswamy, L.D. (1994). *International Environmental Law and World Order.* West Publishing, St. Paul, Minnesota, 1200 pp.

Harmancioglu, N.B., et al.; Editors (1997). *Integrated Approach to Environmental Data Management Systems*. Kluwer Academic Publishers, Boston, Massachusettes, 546 pp.

Hastings, T. (1993). *Proceedings of the Florida Environmental Expo. '93.* Tampa, Florida, October 11 - 13, 1994. Available from FCE, Inc., Clearwater, Florida.

Henderson, T. and Hubler, K.; Editors (1985). *Directory of State Environmental Agencies, 2nd ed.* Environmental Law Institute, Washington, D.C., 273 pp.

Hess, K. (1993). *Environmental Site Assessment: Phase I: A Basic Guide*. Lewis Publishers, Inc., Boca Raton, Florida, 287 pp.

Hill K. and Piccirelli, A.; Editors (1992). *Gale Environmental Sourcebook: A Guide to Organizations, Agencies, and Publications*. Gale Research Inc., Detroit, Mich., 688 pp.

Hinchee, R.F., et al.; Editors (1995). *Applied Bioremediation of Petroleum Hydrocarbons*. Battelle Press, Columbus, Ohio, 550 pp.

Hinchee, R.F., et al.; Editors (1995). *Bioremediation of Chlorinated Solvents*. American Society of Civil Engineers, Reston, Virginia, 350 pp.

Hinchee, R.F.; Editor (1994). *Air Sparging for Site Remediation*. Lewis Publishers, Inc., Boca Raton, Florida 160 pp.

Howard, A. and Remson, I. (1978). *Geology in Environmental Planning.* McGraw-Hill, New York, New York, 478 pp.

Keller, E. (1982). *Environmental Geology*, 3rd ed. Merrill Publishing, Columbus, Ohio, 528 pp.

Kofi Asante-Duah, D. (1997). *Managing Contaminated Sites: Problem Diagnosis and Development of Site Restoration.* John Wiley & Sons, Inc., New York, New York, 260 pp.

Kostecki, P.T. and Calabrese, E.J.; Editors (1990). *Remediation Techniques, Environmental Fate, and Risk Assessment*. Lewis Publishers, Inc., Boca Raton, Florida, 3 vols.

Krismann, C.H. and Shepherd, S.D. (1995). *Environmental Concerns: Directory of Businesses, Organizations, and Government Agencies*. University of Colorado, Business Research Division, Boulder, Colorado, 296 pp.

Lim, C.H., Nolan, G.L., and Wong, J.H.C. (1997). *Design of Remediation Systems*. Lewis Publishers, Inc., Boca Raton, Florida, 263 pp.

Lowe, J. S. and Trauner, T.J., Jr. (1997). *Construction on Contaminated Sites*. McGraw-Hill, New York, New York, 254 pp.

MacDougall, J. (1995). *A Short History of Planet Earth: Mountains, Mammals, Fire, & Ice*. John Wiley & Sons, Inc., New York, New York, 256 pp.

Mackey, H.E., Jr., (1994). Proceedings of the International Nuclear and Hazardous Waste Management Conference. Atlanta, Georgia, August 14 - 18, 1994. Available from American Nuclear Society, Lagrange Park, Illinois.

Mahmood, A., (1991). Contaminated Soil in Urban Construction, *Waste Business West Magazine*, December, 1991, p. 24-25.

Mahmood, A. (1991). Protective Liners used in Seismic Design Projects, *California Consulting Engineer*, Winter 1991, 2 pp.

Mahmood, A., Kang, J. and Shah, T. (1996). Moisture and Vapor Extraction Don't Mix, Cover Article, *Soil and Groundwater Cleanup*, March, 1996, 3pp.

Mann, M.J., et al. (1994). *Innovative Site RemediationTechnology*, Vol. 3: Soil Washing / Soil Flushing. American Academy of Environmental Engineers, Annapolis, Maryland, 160 pp.

Mann, M.J., et al. (1997). *Liquid Extraction Technologies*, Vol. 3: Soil Washing, Soil Flushing American Academy of Environmental Engineers, Annapolis, Maryland, 200 pp.

Mauch, J.C. (1993). *Proceedings of the National Association of Environmental Professionals 18th Annual Conference.* Raleigh, North Carolina, May 24 - 26, 1993. Available from NAEP, Washington, D.C.

McCoy, B.J. and Arulanandan, K. (1991). *Proceedings of the Petrochemical Expo. '91 - 1991 Spring National Meeting.* Houston, Texas, April 7 - 11, 1991. Available from American Institute of Chemical Engineers, New York, New York.

McKenzie, G. (1995). *Investigations in Environmental Geoscience*. Burgess Publishing Co., Minneapolis, Minn., 173 pp.

Means, J.L. and Hinchee, R.E.; Editors (1994). *Emerging Technology for Bioremediation of Metals*. Lewis Publishers, Inc., Boca Raton, Florida, 160 pp.

Meters, R.A., Editor (1997). *Encyclopedia of Environmental Analysis & Remediation*. John Wiley & Sons,

Inc., New York, New York, 8 Vols.

Miller, R.K. and Rupnow, M.E. (1991). *Site Remediation.* Future Tech Surveys Publishing, Survey on Technology and Markets Series, Vol. 150, 50 pp.

Norris, R.D., et al. (1995). *In-Situ Bioremediation of Ground Water and Geologic Material: A Review of Technologies.* DIANE Publishing Co., Upland, Pennsylvania, 200 pp.

Ortolano, L. (1996). *Environmental Regulation and Impact Assessment.* John Wiley & Sons, Inc., New York, New York, 604 pp.

Ostler, (1995). *Environmental Regulations Overview.* Prentice-Hall, Paramus, New Jersey, 288 pp.

Oswald, D. (1997). *Sanborn Fire Insurance Maps: Their History & Applications.* Lacewing Paperbacks, 100 pp.

Patton-Hulce, V.R. (1993). *Environment and the Law: A Dictionary.* ABC-CLIO, Inc., Santa Barbara, California, 361 pp.

Pedersen, T.A. and Curtis, J.T. (1991). *Soil Vapor Extraction Technology.* Noyes Publishing, Pollution Technology Review Series, Vol. 204, 316 pp.

Petts, J., Cairney, T., and Smith, M. (1997). *Contaminated Land Site Investigation and Assessment.* John Wiley & Sons, Inc., New York, New York, 300 pp.

Pollution Engineering Staff (1994). *Bioremediation Desk Manual for the Environmental Professional.* Gulf Publishing Company, Houston, Texas, 97 pp.

Portland Cement Association, (1995). *Potential Solidification / Stabilization Superfund Projects* - 1995 Update. Available from Portland Cement Association, Skokie, Illinois, 48 pp.

Portland Cement Association, (1995). Cement-Based Solidification / Stabilization of Lead-Contaminated Soil at a Utah Highway Construction Site. Reprint of an article published by *Remediation, The Journal of Environmental Cleanup Costs, Technologies & Techniques;* Available from Portland Cement Association, Skokie, Illinois, 8 pp.

Powers, J.P., (1981). *Construction Dewatering: A Guide to Theory and Practice.* John Wiley & Sons, Inc., New York, New York, 484 pp.

Powers, J.P., (1986). *Dewatering: Avoiding Its Unwanted Side Effects.* Available from State Mutual Book and Periodical Service, Ltd., New York, New York, 84 pp.

Powers, J.P., (1992). *Construction Dewatering: New Methods and Applications*, 2nd ed. John Wiley & Sons, Inc., New York, New York, 528 pp.

Reddi, L.N., et al.; Editors (1996). *Non-Aqueous Phase Liquids (Napls) in Subsurface Environment : Assessment and Remediation.* American Society of Civil Engineers, Reston, Virginia

Remediation Technologies, Inc. (1989). *Proceedings of the Management of Manufactured Gas Plant Sites - Technology Transfer Seminar.* Pittsburgh, Pennsylvania, April 19 - 21, 1989. Available from Electric Power Research Institute, Palo Alto, California.

Riser-Roberts, E. (1992). *In-Situ-On-Site Bioremediation of Refined Oils and Fuels.* C.K. Smoley & Sons, a Division of Lewis Publishers, Inc., Boca Raton, Florida.

Rittman, B.E., et al. (1994). *In-Situ Bioremediation.* Noyes Data Corporation, Westwood, New Jersey, 260 pp.

Roberts, D.J. and Roberts, K.A. (1994). *Use of Landfarming to Remediate Soil Contaminated by Pesticides.* Andrews Environmental Engineering, Springfield, Illinois, 53 pp.

Sabatini, D.A. and Knox, R.C.; Editors (1992). *Transport and Remediation of Subsurface Contaminants: Collodial, Interfacial, and Subsurface Phenomena.* American Chemical Society, Washington D.C., 252 pp.

Saxman, D. (1993). *Bioremediation of Hazardous Wastes, Wastewater, and Municipal Waste.* Business Communication Company, Inc., Norwalk, Connecticut, 248 pp.

Seredich, J.; Editor (1990). *Your Research Guide to Environmental Organizations.* Smiling Dolphins Press, Irvine, California, 514 pp.

Sims, M. (1993). *Bioremediation: Nature's Cleanup Tool.* DIANE Publishing Co., Upland, Pennsylvania, 29 pp.

Smith, L.A. (Editor) and Means, J.L. (1995). *Remedial Options for Metals-Contaminated Sites.* Lewis Publishers, Inc., Boca Raton, Florida, 240 pp.

Spain, J.C., et al. (1995). *Innovative Site Remediation Technology,* Vol. 1: Bioremediation. American Academy of Environmental Engineers, Annapolis, Maryland, 175 pp.

Spitz, K. And Moreno, J. (1996). *A Practical Guide to Groundwater and Solute Transport Modeling.* John Wiley & Sons, Inc., New York, New York, 461 pp.

Stecker, S.M. and Pedersen, B.S. (1992). *Proceedings of Environmental Site Assessments: Case Studies and Strategies. Orlando, Florida, August 5 - 7, 1992.* Available from National Ground Water Association, Columbus, Ohio.

Stirling, D.A. and Cho, Y. (1992). *Proceedings of the Hazardous Materials Control Resources Institute 13th Annual National Conference & Exhibition.* Washington, D.C., December 1 - 2, 1992. Available from HMCRI, Greenbelt, Maryland.

Sullivan, T.F.; Editor (1997). *Environmental Law Handbook, 14th ed.* Government Institutes, Inc., Rockville, Maryland, 550 pp.

Tarbuck, E. and Lutgens. F. (1996). *The Earth: an Introduction to Physical Geology*. Prentice Hall, Upper Saddle River, New Jersey, 605 pp.

Thomas, R.J. (1990). *Executive's Guide to Environmental Regulations*. American Somerset, Inc., Bellevue, Washington, 173 pp.

Troxler, W., et al. (1997). *Thermal Desorption*. American Academy of Environmental Engineers, Annapolis, Maryland, 150 pp.

U.S. Environmental Protection Agency, (1994). *Abstracts from the Fifth Annual Symposium on Bioremediation of Hazardous Wastes Research, Development, and Field Evaluations, San Francisco, California, June 28-30-1994*. Available from U.S.E.P.A., Office of Research and Development, Washington, D.C., 119 pp.

[Unknown author / editor] (1994). *Field Applications of Bioremediation*: A Complete Directory. DIANE Publishing Company, Upland, Pennsylvania, 53 pp.

[Unknown author / editor] (1995*). GIS-Key Environmental Data Management System: Innovative Technology Evaluation Report*. DIANE Publishing Company, Upland, Pennsylvania, 131 pp.

[Unknown author / editor] (1995). *Environmental Regulations and Technology*. Gordon Press Publishers, New York, New York.

Verbyla, D.L. and Kang-tsung, C. (1997). *Processing Digital Images in GIS*: A Tutorial Featuring ArcView and Arc/Info. High Mountain Press, Inc., Santa Fe, New Mexico, 418 pp.

Vidic R.D., and Pohland, F.; Editors (1995*). Innovative Technologies for Site Remediation and Hazardous Waste Management*. Proceedings of the National Conference, Pittsburgh, Pennsylvania, July 23-26, 1995. American Society of Civil Engineers, Reston, Virginia.

Ward, C.H. (1994). *Innovative Site Remediation Technology: Bioremediation*; Vol I. American Academy of Environmental Engineers, Annapolis, Maryland.

Weitzman, L., Jefcoat I.A., and Kim, B.R.; Editors (1997*). Innovative Site Remediation Technology: Design and Application.* American Academy of Environmental Engineers, Annapolis, Maryland.

Wildavsky, A. (1997). *But Is It True: A Citizens Guide Environmental Health and Safety Issues*. Harvard University Press, Cambridge, Massachusettes.

Wiley, S. and Taylor, L. (1995). *Proceedings of the Second World Congress of the Society of Environmental Toxicology and Chemistry.* Vancouver, British Columbia, November 5 - 9, 1995. Available from SETAC Press, Pensacola, Florida.

Wilson, D.J. and Clarke, A.N.; Editors (1994*). Hazardous Waste Site Soil Remediation: Theory and Application of Innovative Technologies*. Marcel Dekker, Inc., New York, New York, 584 pp.

Woodside, G. and Kocurek, D. (1997). *Environmental Safety and Health Engineering*. John Wiley & Sons,

Inc., New York, New York.

Yuhr, L. (1994). *Proceedings of the Florida Environmental Expo. '94*. Tampa, Florida, October 11 - 13, 1994. Available from FCE, Inc., Clearwater, Florida.

Zeiler, M. (1997). *Inside ARC/INFO, Revised Ed.* High Mountain Press, Inc., Santa Fe, New Mexico, 640 pp.

INDEX

GOVERNMENT INSTITUTES ORDER FORM

4 Research Place, Suite 200 • Rockville, MD 20850-3226 • Tel (301) 921-2323 • Fax (301) 921-0264
Internet: *http://www.govinst.com* • E-mail: *giinfo@govinst.com*

3 EASY WAYS TO ORDER

1. Phone: **(301) 921-2323**
Have your credit card ready when you call.

2. Fax: **(301) 921-0264**
Fax this completed order form with your company purchase order or credit card information.

3. Mail: **Government Institutes**
4 Research Place, Suite 200
Rockville, MD 20850-3226
USA
Mail this completed order form with a check, company purchase order, or credit card information.

PAYMENT OPTIONS

❑ **Check** (*payable to Government Institutes in US dollars)*

❑ **Purchase Order** (this order form must be attached to your company P.O. Note: All International orders must be pre-paid.)

❑ **Credit Card** ☐ VISA ☐ MasterCard ☐ American Express

Exp. ___/___

Credit Card No. ____________________

Signature ____________________

Government Institutes' Federal I.D.# is 52-0994196

CUSTOMER INFORMATION

Ship To: (Please attach your Purchase Order)

Name: ____________________

GI Account# (*7 digits on mailing label*): ____________________

Company/Institution: ____________________

Address: ____________________
(please supply street address for UPS shipping)

City: __________ State/Province: __________

Zip/Postal Code: __________ Country: __________

Tel: ( ) ____________________

Fax: ( ) ____________________

E-mail Address: ____________________

Bill To: (if different than ship to address)

Name: ____________________

Title/Position: ____________________

Company/Institution: ____________________

Address: ____________________
(please supply street address for UPS shipping)

City: __________ State/Province: __________

Zip/Postal Code: __________ Country: __________

Tel: ( ) ____________________

Fax: ( ) ____________________

E-mail Address: ____________________

Qty.	Product Code	Title	Price

❑ **New Edition No Obligation Standing Order Program**

Please enroll me in this program for the products I have ordered. Government Institutes will notify me of new editions by sending me an invoice. I understand that there is no obligation to purchase the product. This invoice is simply my reminder that a new edition has been released.

Subtotal __________

MD Residents add 5% Sales Tax __________

Shipping and Handling (see box below) __________

Total Payment Enclosed __________

Within U.S:	**Outside U.S:**
1-4 products: $6/product	Add $15 for each item (Airmail)
5 or more: $3/product	Add $10 for each item (Surface)

15 DAY MONEY-BACK GUARANTEE

If you're not completely satisfied with any product, return it undamaged within 15 days for a full and immediate refund on the price of the product.

SOURCE CODE: BP01